Applied Intelligence in Human–Computer Interaction

The text comprehensively discusses the fundamental aspects of human–computer interaction, and applications of artificial intelligence in diverse areas including disaster management, smart infrastructures, and healthcare. It employs a solution-based approach in which recent methods and algorithms are used for identifying solutions to real-life problems.

This book:

- Discusses the application of artificial intelligence in the areas of user interface development, computing power analysis, and data management.
- Uses recent methods/algorithms to present solution-based approaches to real-life problems in different sectors.
- Showcases the applications of artificial intelligence and automation techniques to respond to disaster situations.
- Covers important topics such as smart intelligence learning, interactive multimedia systems, and modern communication systems.
- Highlights the importance of artificial intelligence for smart industrial automation and systems intelligence.

The book elaborates on the application of artificial intelligence in user interface development, computing power analysis, and data management. It explores the use of human–computer interaction for intelligence signal and image processing techniques. The text covers important concepts such as modern communication systems, smart industrial automation, interactive multimedia systems, and machine learning interface for the internet of things. It will serve as an ideal text for senior undergraduates, and graduate students in the fields of electrical engineering, electronics and communication engineering, computer engineering, and information technology.

Applied Intelligence in Human–Computer Interaction

Edited by
Sulabh Bansal
Prakash Chandra Sharma
Abhishek Sharma
Jieh-Ren Chang

CRC Press
Taylor & Francis Group
Boca Raton London New York

CRC Press is an imprint of the
Taylor & Francis Group, an **informa** business

First edition published 2023
by CRC Press
6000 Broken Sound Parkway NW, Suite 300, Boca Raton, FL 33487-2742

and by CRC Press
4 Park Square, Milton Park, Abingdon, Oxon, OX14 4RN

CRC Press is an imprint of Taylor & Francis Group, LLC

ISBN: 978-1-032-39276-9 (hbk)
ISBN: 978-1-032-54165-5 (pbk)
ISBN: 978-1-003-41546-6 (ebk)

DOI: 10.1201/9781003415466

Typeset in Sabon
by SPi Technologies India Pvt Ltd (Straive)

Contents

Editors

Jieh-Ren Chang earned his BS and MS degrees in nuclear science engineering from National Tsing Hua University, Taiwan, in 1986 and 1988, respectively. He is currently a professor in the Department of Electronic Engineering, National Ilan University (NIU), Taiwan. He is the Dean of the Library and Information Center and leads the Intelligent Information Retrieval Laboratory at NIU. Prior to joining NIU, he worked as a technician at the Missile Manufacturing Institute of the Sun Yat-sen Academy of Sciences under the Ministry of National Defense of Taiwan. He has also worked for System Studies & Simulation, Inc., a real-time simulation company in the USA. He has created several software systems for different government organizations and industrial companies. His current research interests are in image processing, knowledge discovery from databases and the Internet, machine deep learning and data mining, embedded systems, AIOT systems in industrial manufacturing processes, and innovative combinations of information and communication technologies.

Abhishek Sharma earned his B.E, in Electronics Engineering from Jiwaji University, Gwalior, India, and Ph.D. in Engineering from the University of Genoa, Italy. He is presently working as an associate professor in the Department of Electronics and Communication Engineering at The LNM Institute of Information and Technology, Jaipur, RJ, India. He is also a member of IEEE, Computer Society, and Consumer Electronics Society, a Lifetime member of Indian Society for Technical Education (ISTE), a Lifetime member of Advanced Computing Society (ACS), India, and a Life Member of Computer Society of India (CSI). He is also the coordinator of the Texas Instruments Lab, and Intel Intelligent Lab in the present Institute. He is also the center lead of the LNMIIT-Centre of Smart Technology (L-CST). He has authored and edited books, his research work is published in reputed conferences and journals. His current research interests are Real-Time Systems, Embedded Systems, Computer Vision, and IoT.

Sulabh Bansal is an associate professor at Manipal University, Jaipur. He has 22 years experience in the field of information technology and computer science, working on software application development, and research and academic projects. He earned his BEng from Shivaji University Kolhapur in 2000, going on to complete an MTech in 2012 and a PhD in algorithms for optimization problems at Dayalbagh Educational Institute in 2017. He is a Senior Member of the IEEE, a lifetime member of the Indian Society for Technical Education (ISTE), and a member of the Association for Computing Machinery India Council. He has published several research papers in SCI- and Scopus-indexed journals, presented his work at different international conferences, and delivered keynote addresses at various national and international academic events. He has supervised a number of BTech and MTech projects, and PhD students. His research interests include evolutionary algorithms, optimization problems, quantum-inspired computing, and machine learning.

Prakash Chandra Sharma is an associate professor in the School of Information Technology at Manipal University, Jaipur, India. He earned his Master of Engineering in computer engineering from SGSITS Indore, and his PhD in computer science and engineering from the Indian Institute of Technology, Indore. He has more than 16 years' teaching and research experience in various areas of computer science and engineering. His research interests include graph theory, theoretical computer science, artificial intelligence, image processing and soft computing. He has taught many subjects at undergraduate and postgraduate level, including data structures and algorithms, theory of computation, operating systems, compiler design, artificial intelligence, semantic web, graph theory, soft computing, and discrete mathematics. He has published a number of book chapters, as well as research papers in various SCI/Scopus-indexed international journals and in proceedings of national and international conferences. He is an active reviewer and editorial member of various reputed international journals in his research areas. He is a member of various international professional associations, including IEEE, CSI, IAENG, and the Institute of Engineers India. He has received many national awards.

Contributors

Punit Agarwal
The LNM Institute of Information
  Technology,
Jaipur, Rajasthan, India

Abhigyan Agarwala
The LNM Institute of Information
  Technology, Jaipur, Rajasthan,
  India

R Ahila Priyadharshini
Centre for Image Processing and
  Pattern Recognition, Mepco
  Schlenk Engineering College,
  Sivakasi, India

S Arivazhagan
Centre for Image Processing and
  Pattern Recognition, Mepco
  Schlenk Engineering College,
  Sivakasi, India

M Arun
Centre for Image Processing and
  Pattern Recognition, Mepco
  Schlenk Engineering College,
  Sivakasi, India

Vikas Bajpai
The LNM Institute of Information
  Technology, Jaipur, Rajasthan,
  India

Anukriti Bansal
The LNM Institute of Information
  Technology, Jaipur, Rajasthan,
  India

Sulabh Bansal
Department of Information
  Technology
School of Information
  Technology
Manipal University Jaipur, Jaipur,
  Rajasthan, India

Shiladitya Bhattacharjee
Informatics Cluster,
School of Computer Science,
UPES, Dehradun, Uttarakhand,
  India

Pratyush Chahar
The LNM Institute of Information
  Technology,
Jaipur, Rajasthan, India

Wei-Chih Chen
National Yang Ming Chiao Tung
  University
Hsinchu, Taiwan

Khushboo Dadhich
JK Lakshmipat University
Jaipur, Rajasthan, India

Vibhav Garg
The LNM Institute of Information
 Technology,
Jaipur, Rajasthan, India

Amrata Gill
Faculty of Engineering,
Dayalbagh Educational Institute,
 Agra, India

Armaan Jain
Chitkara University Institute of
 Engineering and Technology,
Patiala, Punjab, India

Meet Kumar Jain
The LNM Institute of Information
 Technology,
Jaipur, Rajasthan, India

Devika Kataria
JK Lakshmipat University,
Jaipur, Rajasthan, India

Arshpreet Kaur
Department of Computer Science
 and Engineering, Alliance College
 of Engineering & Design,
Alliance University, Bangalore,
 Karnataka, India

Munish Khanna
Department of Computer Science
 and Engineering,
Hindustan College of Science and
 Technology, Mathura, India

Vinay Kukreja
Chitkara University Institute of
 Engineering and Technology,
 Chitkara University Rajpura,
 Punjab, India

Deepak Kumar
Chitkara University Institute of
 Engineering and Technology,
 Chitkara University Rajpura,
 Punjab, India

Reetu Malhotra
Chitkara University Institute of
 Engineering and Technology,
 Chitkara University, Punjab
Patiala, Punjab, India

Ashray Mittal
The LNM Institute of Information
 Technology,
Jaipur, Rajasthan, India

Adarsh Pandey
Cadence Design Systems (India)
 Pvt. Ltd., Noida, India

Siby Samuel
Department of Computer Science,
 St. Aloysius (Autonomous)
 College,
Jabalpur, India

Katakam Venkata Seetharam
Department of Civil Engineering,
 Sreenidhi Institute of Science and
 Technology,
Yarnnampet, Ghatkesar,
 Hyderabad, India

Abhishek Sharma
The LNM Institute of Information
 Technology
Jaipur, India

Kartikey Sharma
The LNM Institute of Information
 Technology, Jaipur, Rajasthan,
 India

Prakash Chandra Sharma
Department of Information
 Technology, Manipal University
 Jaipur
Jaipur, Rajasthan, India

Vinay Kumar Sharma
The LNM Institute of Information
 Technology,
Jaipur, Rajasthan, India

Kumar Shashvat
Department of Computer Science
  and Engineering, Alliance College
  of Engineering & Design,
Alliance University, Bangalore,
  Karnataka, India

Law Kumar Singh
Department of Computer
  Engineering and Applications
GLA University, Mathura, India

Lokesh Singh
Department of Computer Science
  and Engineering, Alliance College
  of Engineering & Design,
Alliance University,
Bangalore, Karnataka, India

Rekha Singh
Department of Physics,
U.P.Rajarshi Tandon Open University,
Prayagraj, India

Tarun Singh
The LNM Institute of Information
  Technology,
Jaipur, Rajasthan, India

Deepti Sisodia
Department of Computer Science
  and Engineering, National
  Institute of Technology,
Raipur, Chhattisgarh, India

Nidhip Taneja
Keysight Technologies, Gurugram,
  India

Stephen Vincent
The LNM Institute of Information
  Technology,
Jaipur, Rajasthan, India

Prem Prakash Vuppuluri
Department of Electrical
  Engineering, Faculty of
  Engineering,
Dayalbagh Educational Institute
  (Deemed University), Agra,
  India

Hwang-Cheng Wang
National Ilan University
Yilan, Taiwan

Shreyash Yadav
Department of Electronics
  and Communication
  Engineering,
The LNM Institute of Information
  Technology,
Jaipur, Rajasthan, India

Chapter 1

Prediction of extreme rainfall events in Rajasthan

*Vikas Bajpai, Anukriti Bansal, Stephen Vincent,
Abhigyan Agarwala and Kartikey Sharma*

The LNM Institute of Information Technology, Mukandpura, India

CONTENTS

1.1 INTRODUCTION

This chapter deals with the use of time-series rainfall data to predict extreme rainfall events and discover the features that influence rainfall. Advance prediction of such events can prevent the loss of economic resources and mitigate the threat to human lives. The dataset suffers from imbalance, with most instances being scanty or no rainfall events. Long Short-term Memory (LSTM) networks were able to perform better than other neural networks and decision trees. Using the LSTM network, we predict the next day's rainfall while considering the weather parameters for varying timesteps. However, the networks generally fail to predict the test data's rainfall accurately. The data imbalance prevents the learning of attributes for moderate and heavy rainfall, and the model tends to be biased towards instances of scanty rainfall. Time-series GAN or SMOTE-Regression methods could possibly reduce this imbalance by oversampling the heavy rainfall observations.

DOI: 10.1201/9781003415466-1

1.1.1 The area of work

Most real-world cases involve time-dependent datasets. These include climate predictions, language translators, and voice assistants like Siri and Cortana. All possible decisions are taken not only on the basis of current interactions or instances but also past ones. Time-series data is different from ordinary data in the sense that the learning model needs to retain past information in order to make accurate predictions. Since most of the machine learning models do not take temporal relation between observations into account, they are not suitable for time-dependent predictions. Recurrent neural network (RNN) is a feedforward network with an internal memory. It takes the current input and previous output for computation. The output generated is fed back to the network to introduce time dependency. RNNs suffer from the vanishing gradient problem, which means that as the learning progresses, it is difficult for the model to retain past information. This is due to diminishing gradients during backward propagation through time. To overcome this issue, LSTM [1] nodes are used. These nodes have gates to control which information to keep and which information to discard for better retention and prediction.

1.1.2 Problem addressed

The problem we have addressed is predicting heavy rainfall events over the state of Rajasthan. Prediction of heavy rainfall in advance is vital for meteorologists, as it prevents loss of human lives in case of flooding and mitigates damage to various weather-dependent activities, agriculture, and hydrology. Predicting heavy rainfall has remained a challenging task due to the non-Gaussian distribution of rainfall, and the recorded data being highly skewed, with few instances of heavy rainfall.

1.1.3 Literature survey

There have been various works on rainfall prediction in India [2, 3]. In the majority of them heavy rainfall has been related to anomalous weather behavior. Though there has been success in predicting rainfall in general, predicting heavy rainfall accurately in advance remains a challenge.

In one work, Gope et al. [4] explored deep learning with SAE to learn and represent weather features and use them to predict extreme precipitation events 24–48 hours in advance. Subrahmanyam et al. [5] applied Gaussian process regression to time-series rainfall data to predict extreme rainfall events, and had promising results. It was shown in [6] that the Deep Echo State Network (DeepESN) model could be a reliable tool to predict rainfall events. They also show that rainfall, humidity and pressure are crucial factors in predicting rainfall.

In a recently published study by Hess et al. [7], the researchers applied deep neural networks (DNNs) to correct biases in the rainfall forecast of a numerical weather prediction (NWP) ensemble. They optimized the DNN with a weighted loss function to account for heavy rainfall events, showing significant improvement in prediction performance.

In yet another work, Ravuri et al. [8] tackle the problem of predicting moderate-to-heavy rainfall using deep generative models; they provide promising results where most other models struggle, improving forecast quality, consistency and value.

These studies noted the lack of accurate verification metrics for predicted values, highlighting the need for better quantitative measures which would also align with operational utility to evaluate models. With recently changing climatic patterns, it is expected that there will be an increase in extreme rainfall events, according to Goswami et al. [9]. This makes it important to have a thorough scientific understanding of rainfall extremes, for accurate predictions.

There have not been many advances when it comes to employing deep learning in meteorology. Liu et al. [10] developed a deep neural network model to predict specific environmental parameters over the course of a few hours. Alex Graves [11] employed deep neural networks and proposed a probabilistic graphical model for forecasting various weather parameters. Turning to studies specifically on Indian rainfall, Sahai et al. [3] analyze results by employing artificial neural networks on Indian rainfall data, specifically the Parthasarathy dataset. However, none of them aid in predicting extreme rainfall events, which is paramount when it comes to impacting people's lives.

1.1.4 Dataset creation

We use data from the NCEP/NCAR Reanalysis, which comprises 4-times daily, daily, and monthly values of 65 environmental parameters, across 17 pressure levels and 28 sigma levels, from January 1, 1948 to the present time. Days of only June, July, August and September (being the months having most rainfall) were considered for rainfall prediction. All values were normalized along each column, and missing values were imputed with mean values. We work with the data across six geographical coordinates:

- 27.5 deg North and 70.0 deg East
- 27.5 deg North and 72.5 deg East
- 27.5 deg North and 75.0 deg East
- 25.0 deg North and 70.0 deg East
- 25.0 deg North and 72.5 deg East
- 25.0 deg North and 75.0 deg East

Initially, we will focus on predicting rainfall only for one coordinate, i.e., 27.5 degrees North and 72.5 degrees East. Subsequently, we will include geographical coordinates as input features for certain models. Observing the properties of our newly created dataset, we find that there is a large imbalance in rainfall distribution, i.e., the days having heavy rainfall (and hence rich in information) are very scarce across the dataset, as shown in Figure 1.1.

Figure 1.1 Rainfall (in mm) vs. frequency of days having said rainfall.

1.1.5 Feature importance

Feature importance is the analysis of input features; scores are assigned to features based on how vital they are for predicting a target output. In the context of Random Forests, the importance of each input feature is computed from how 'pure' the leaves of the trees are. We use sklearn's inbuilt functionalities to extract feature importances from a Random Forest Regressor (which we will employ next). We then order the features in decreasing order of their importance, and compute which subset of features comprising the top K features give the best results, which is visualized in Figure 1.2. Finally, we create another dataset based on the 19 features specified in Gope et al. [4], and compare its minimum squared error (MSE) with other datasets. We observe that this dataset yields a better MSE of 28.87 (as shown in Figure 1.3) than any of the subset of features created, taking feature importance into account (the least MSE among them being 29.23). Henceforth,

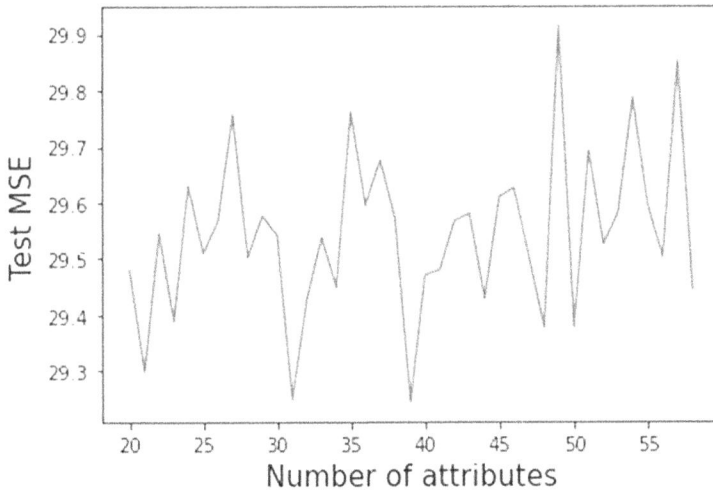

Figure 1.2 Number of features included vs. minimum squared error (MSE).

```
from sklearn.metrics import mean_squared_error
print("training mean squared error : ", mean_squared_error(y_train, y_pred_for_train))
print("testing mean squared error : ", mean_squared_error(y_test, y_pred))

training mean squared error :  6.13998714380947
testing mean squared error :  28.875556805602997
```

Figure 1.3 MSE of 19 features.

we use two datasets for our models: one dataset contains all 65 features, whereas the other dataset comprises the 19 features that yielded the least MSE during feature importance analysis.

1.1.6 Random Forest Regressor

Our first approach to rainfall prediction was employing a Random Forest Regressor. The number of trees were determined using GridSearch. The accuracy of the model was estimated using MSE. The outputs for dataset 1 and 2 are visualized in Figures 1.4 and 1.5 respectively.

1.1.7 Multi-layer perceptron

For the first dataset (65 features), we use a multi-layer perceptron (MLP) having three hidden layers with 128, 10 and 4 nodes. The layer description for dataset 1 is shown in Figure 1.6, while the actual and predicted values are visualized in Figure 1.7.

For the second dataset (19 features), we have three hidden layers with 40, 16 and 4 nodes. The layer description for dataset 2 is shown in Figure 1.8, while the actual and predicted values are visualized in Figure 1.9.

We run both models for 300 epochs (post 300 epochs, no significant improvements in results are observed). The optimal number of layers and nodes in each layer were estimated using trial and error. By observing the plots generated by Random Forest Regressor as well as multi-layer perceptron, we conclude that both models fail to gage an accurate measure of rainfall during days when there was considerable rainfall.

1.1.8 Basic RNN

From here on out, as the models to be employed will be computationally demanding and time-consuming, we will only work with the smaller dataset of 19 features. However, instead of normalizing each feature as we did previously, we scale each feature from 0 to 100 independently. RNNs were used with varying timesteps (days prior). When it comes to training data, we observe that smaller timesteps (3, 4, 5,...) yielded much better results than larger timesteps. Even so, it is evident from the training/testing plots that the model overfits to the training data. The RNN layer description is given in Figure 1.10, and the training loss vs. epoch plot is shown in Figure 1.11.

The actual and predicted values for training and testing data are plotted in Figures 1.12 and 1.13.

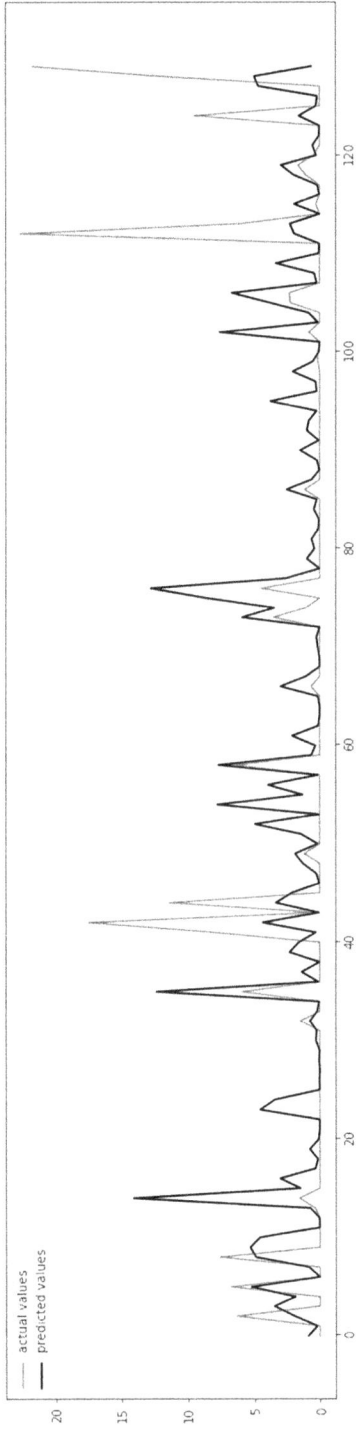

Figure 1.4 Actual vs. predicted values for dataset I.

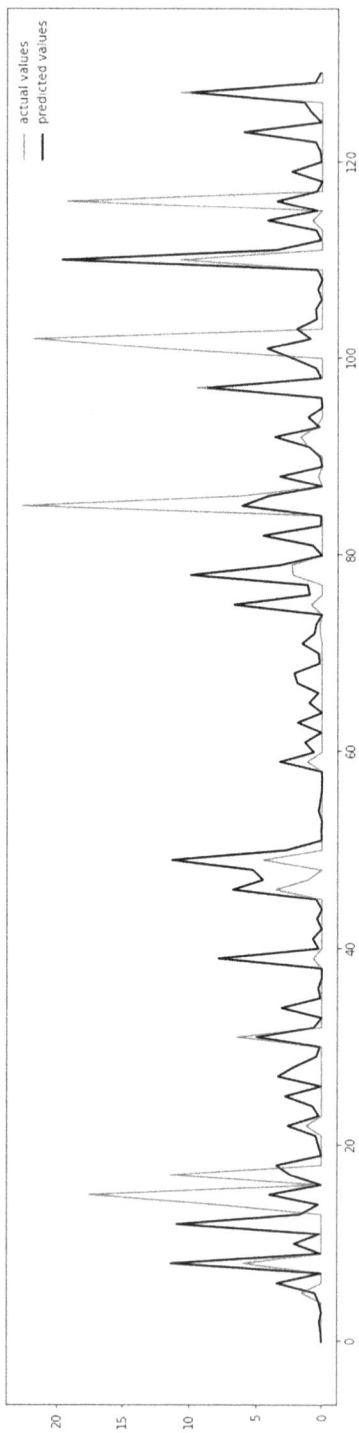

Figure 1.5 Actual vs. predicted values for dataset 2.

```
Model: "sequential_8"

_____
Layer (type)                   Output Shape              Param #
=================================================================
sequential_9 (Sequential)      (None, 128)               8960

_____
sequential_10 (Sequential)     (None, 10)                1330

_____
sequential_11 (Sequential)     (None, 4)                 60

_____
dense_11 (Dense)               (None, 1)                 5
=================================================================
Total params: 10,355
Trainable params: 10,071
Non-trainable params: 284

_____
None
```

Figure 1.6 Layer description of MLP for dataset 1.

1.1.9 LSTM

We now improve upon our results using the LSTM model [1]. LSTM's improved learning allows the user to train models using sequences with several hundred time steps, something the RNN struggles to do. For this model, instead of predicting rainfall for a specific coordinate, we now include the geographical coordinates as input parameters. There are three LSTM BatchNorm blocks with no dropout layer and 256, 128 and 32 nodes respectively. 5-day timesteps are considered, and the model is trained for 250 epochs with a batch size of 32. The LSTM layer description is given in Figure 1.14, and the training loss vs. epoch plot is shown in Figure 1.15.

The batch normalization [12] layer normalizes the attributes of the entire batch of hidden values to have zero mean and unit variance. During training changes in weights and bias can cause change distribution of hidden activation values (internal covariate shift) and it takes longer to train the model. Batch normalization prevents the shift and increases the speed and stability of the model. The actual and predicted values for training and testing data are plotted in Figures 1.16 and 1.17.

We observe that by using 2 days as the timestep instead of 5 days, we have more accurate predictions on training data but minimal improvement for test data, overfitting being a major issue. The model uses four LSTM BatchNorm blocks with 250, 150, 100 and 50 nodes respectively. The model is trained for 250 epochs with a batch size of 64. The LSTM layer description is given in Figure 1.18, and the training loss vs. epoch plot is shown in Figure 1.19.

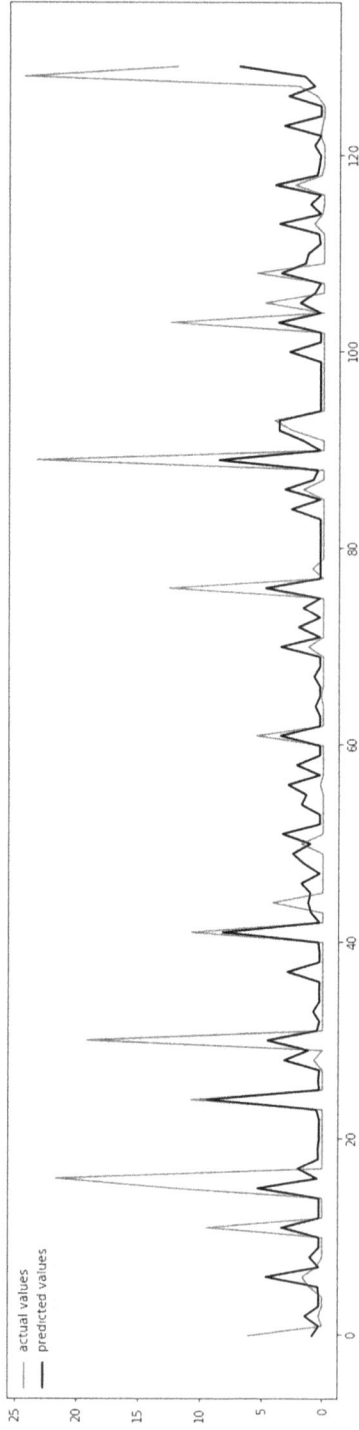

Figure 1.7 Actual vs. predicted values for dataset 1.

```
Model: "sequential"

_____
Layer (type)                 Output Shape              Param #
=================================================================
sequential_1 (Sequential)    (None, 40)                880

_____
sequential_2 (Sequential)    (None, 16)                720

_____
sequential_3 (Sequential)    (None, 4)                 84

_____
dense_3 (Dense)              (None, 1)                 5

=================================================================
Total params: 1,689
Trainable params: 1,569
Non-trainable params: 120

_____
None
```

Figure 1.8 Layer description of MLP for dataset 2.

The actual and predicted values for training and testing data are plotted in Figures 1.20 and 1.21.

On increasing the number of epochs from 250 to 350, the model grossly overfits to the training data and performs even more poorly on testing data. The training and testing data plots are shown in Figure 1.22 and 1.23.

Finally, a timestep of 1 day was used in place of 2 days. In this case, the model even fails to do well at training data. The model employed is similar to the previous one, with 350 epochs and batch size of 64. The training and testing data plots are shown in Figures 1.24 and 1.25.

1.1.10 Conclusion and future work

RNN with a timestep of 2 days gives the most accurate result on the training set compared to other models and other timesteps. However, on the test set all the models perform poorly. This is mainly due to huge imbalance in the data set. The instances of scanty rainfall are much more numerous than heavy rainfall instances. To deal with this issue we can use an oversampling method in regression problems like SMOTE-R (Synthetic Minority Oversampling Technique for Regression) or Time-Series Generative Adversarial Networks (GAN). The key feature of SMOTE-R [13] is that it combines undersampling of frequent classes with oversampling of minority classes. The proposed method is based on a relevance function and on a user-specified threshold for relevance values that leads to a set. SMOTE-R then

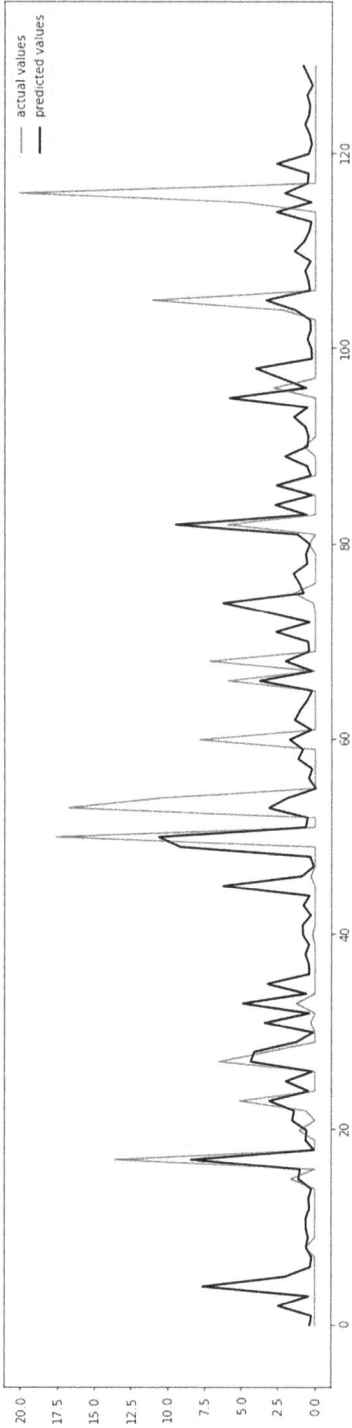

Figure 1.9 Actual vs. predicted values for dataset 2.

```
Model: "sequential"

_____
Layer (type)                 Output Shape              Param #
=================================================================
sequential_1 (Sequential)    (None, 5, 256)            281600

_____
sequential_2 (Sequential)    (None, 5, 128)            197632

_____
sequential_3 (Sequential)    (None, 32)                20736

_____
dense (Dense)                (None, 1)                 33
=================================================================
Total params: 500,001
Trainable params: 499,169
Non-trainable params: 832
_____
None
```

Figure 1.10 RNN layers description.

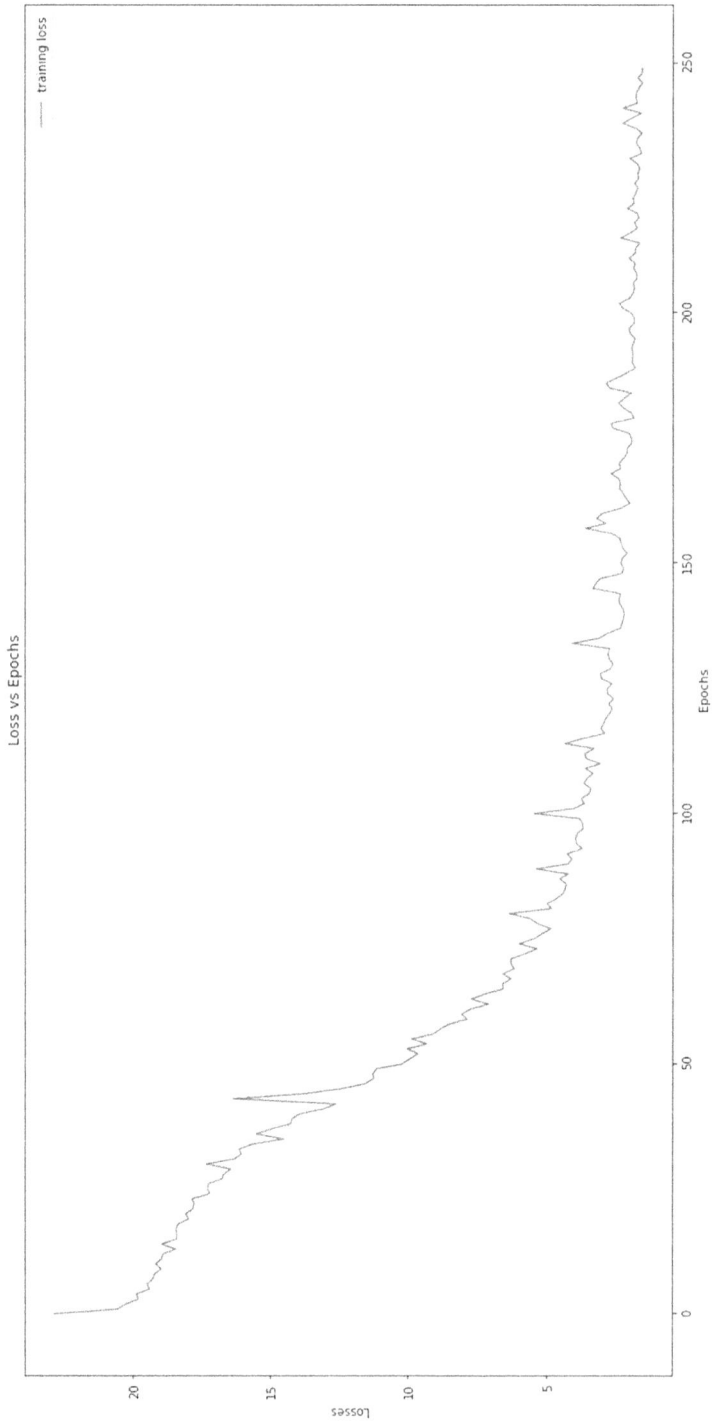

Figure 1.11 Training loss vs. number of epochs.

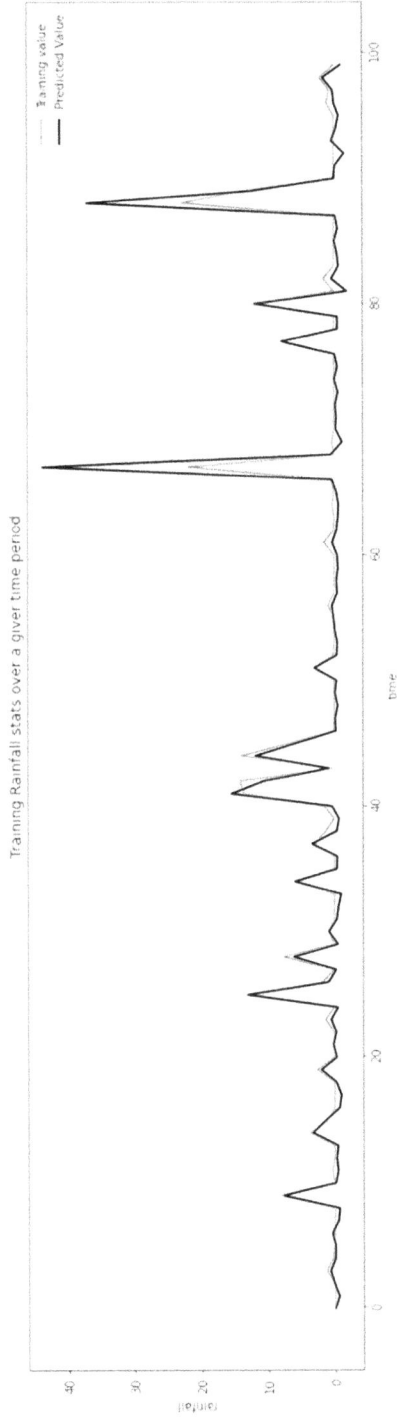

Figure 1.12 Actual vs. predicted values for training data.

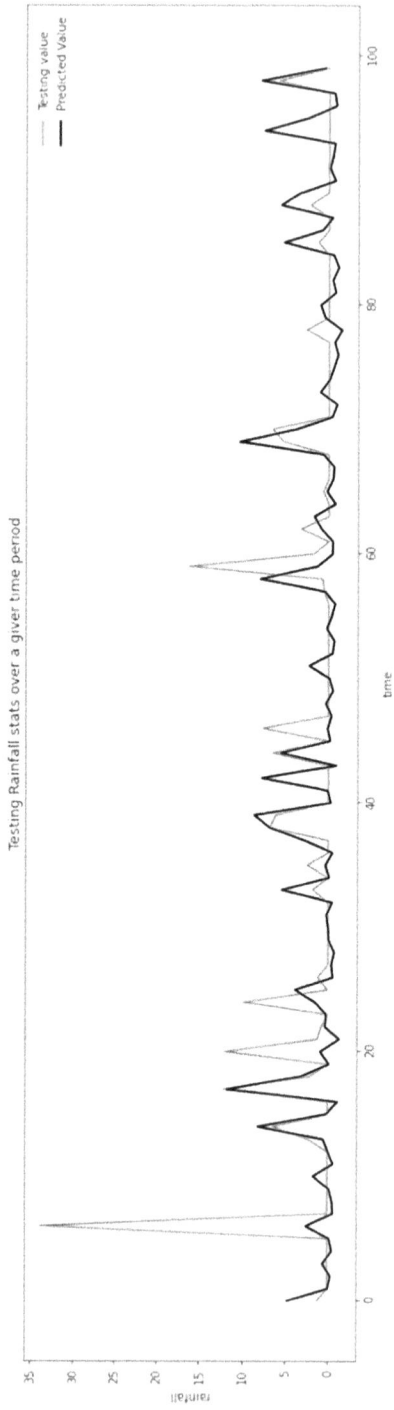

Figure 1.13 Actual vs. predicted values for testing data.

```
Model: "sequential"

_____
Layer (type)                Output Shape              Param #
=================================================================
sequential_1 (Sequential)   (None, 5, 256)            283648

_____
sequential_2 (Sequential)   (None, 5, 128)            197632

_____
sequential_3 (Sequential)   (None, 32)                20736

_____
dense (Dense)               (None, 1)                 33
=================================================================
Total params: 502,049
Trainable params: 501,217
Non-trainable params: 832

_____
None
```

Figure 1.14 LSTM layers description.

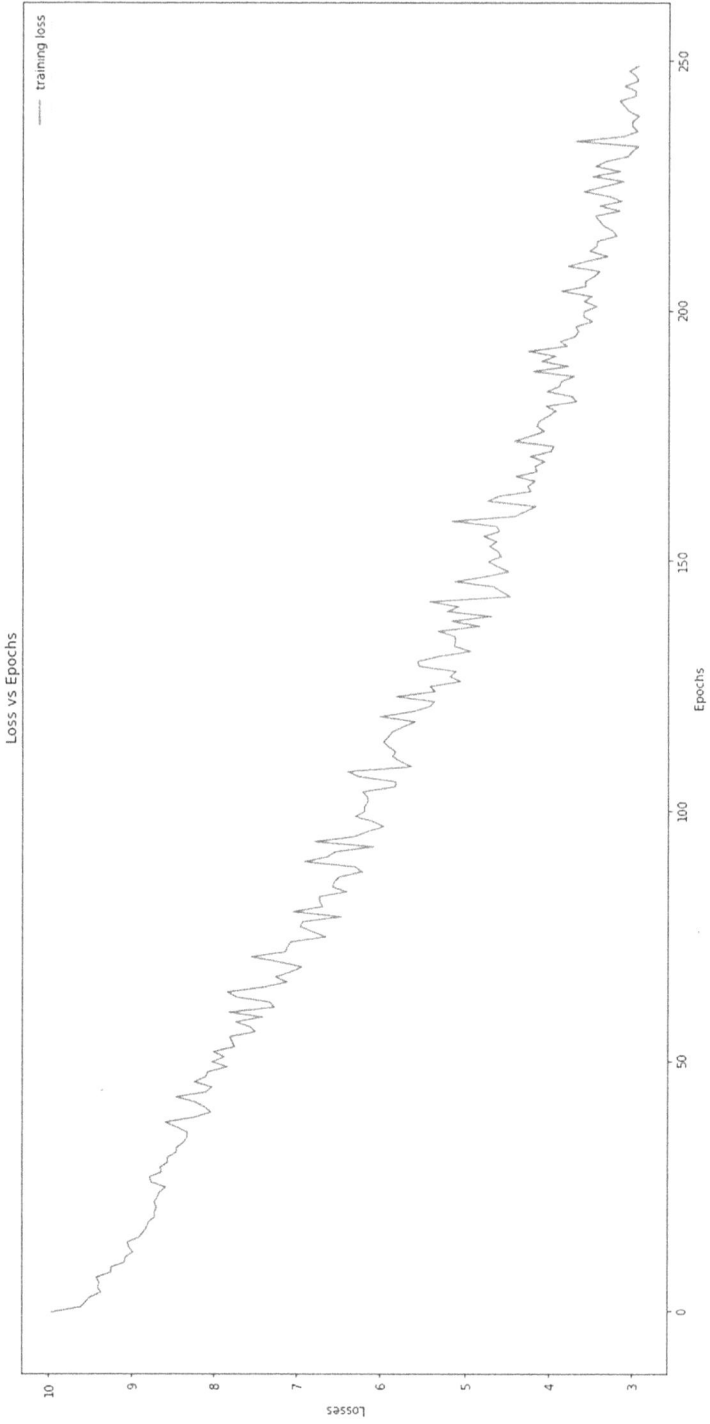

Figure 1.15 Training loss vs. number of epochs.

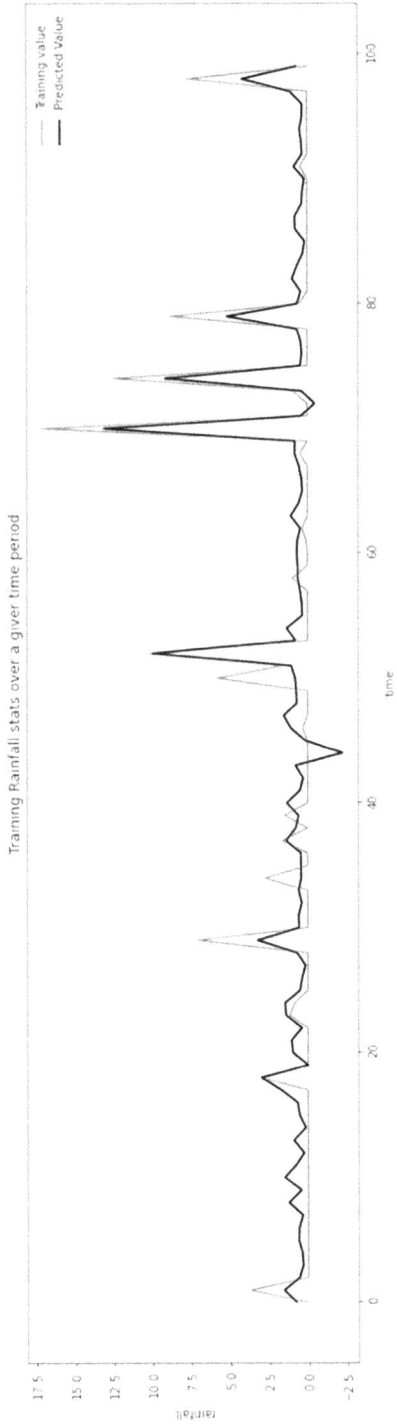

Figure 1.16 Actual vs. predicted values for training data.

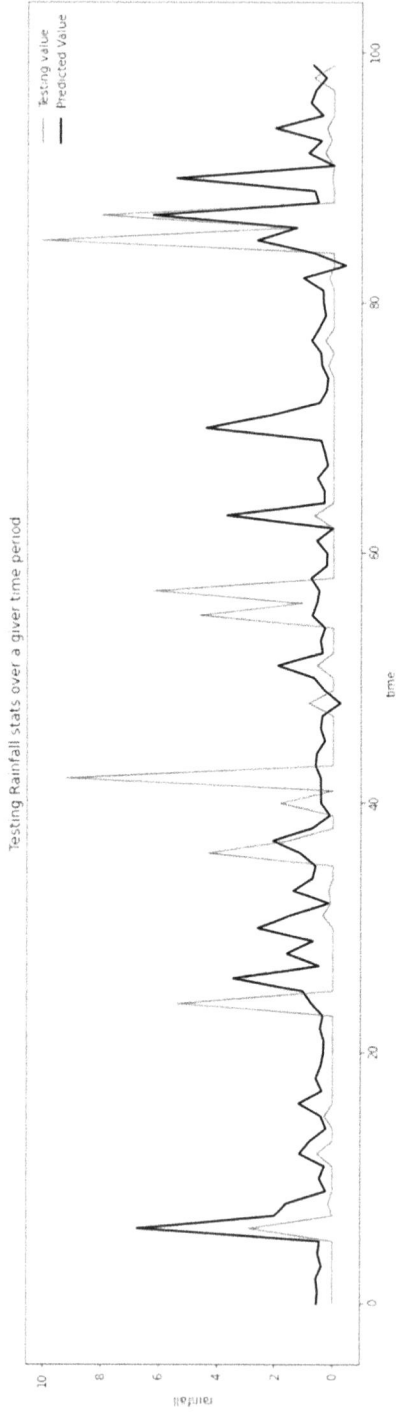

Figure 1.17 Actual vs. predicted values for testing data.

```
Model: "sequential"

_____
Layer (type)                 Output Shape              Param #
=================================================================
sequential_1 (Sequential)    (None, 2, 250)            271000
_____
sequential_2 (Sequential)    (None, 2, 150)            241200
_____
sequential_3 (Sequential)    (None, 2, 100)            100800
_____
lstm_3 (LSTM)                (None, 50)                30200
_____
batch_normalization_3 (Batch (None, 50)                200
_____
dense (Dense)                (None, 1)                 51
=================================================================
Total params: 643,451
Trainable params: 642,351
Non-trainable params: 1,100
_____
None
```

Figure 1.18 LSTM layers description.

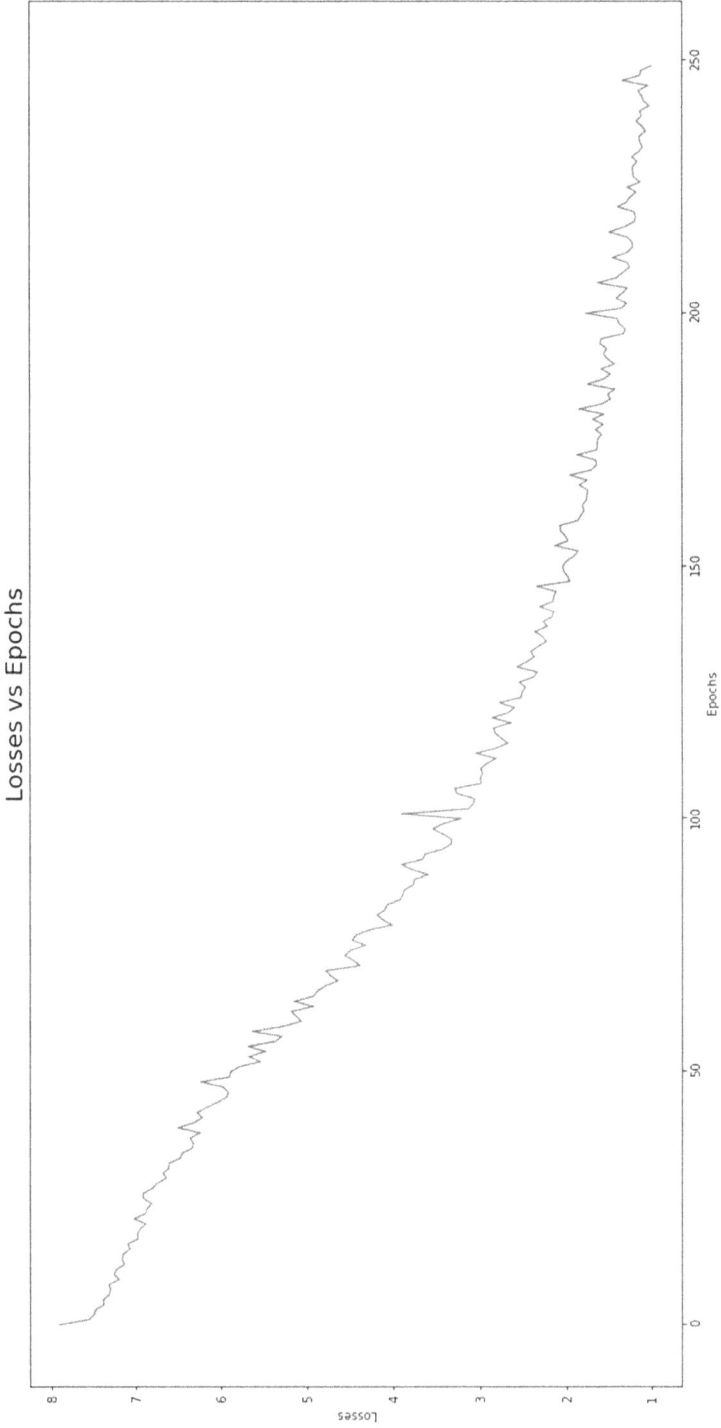

Figure 1.19 Training loss vs. number of epochs.

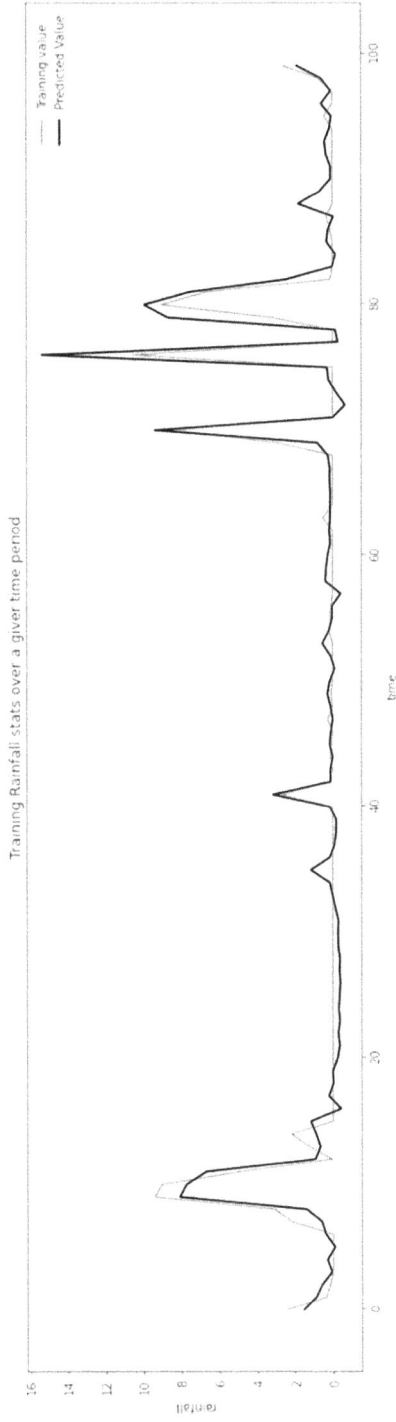

Figure 1.20 Actual vs. predicted values for training data.

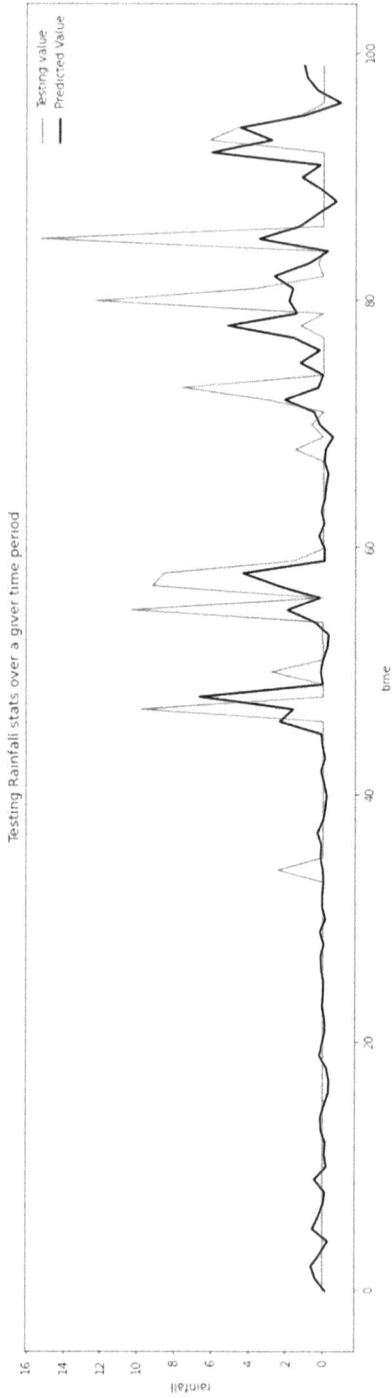

Figure 1.21 Actual vs. predicted values for testing data.

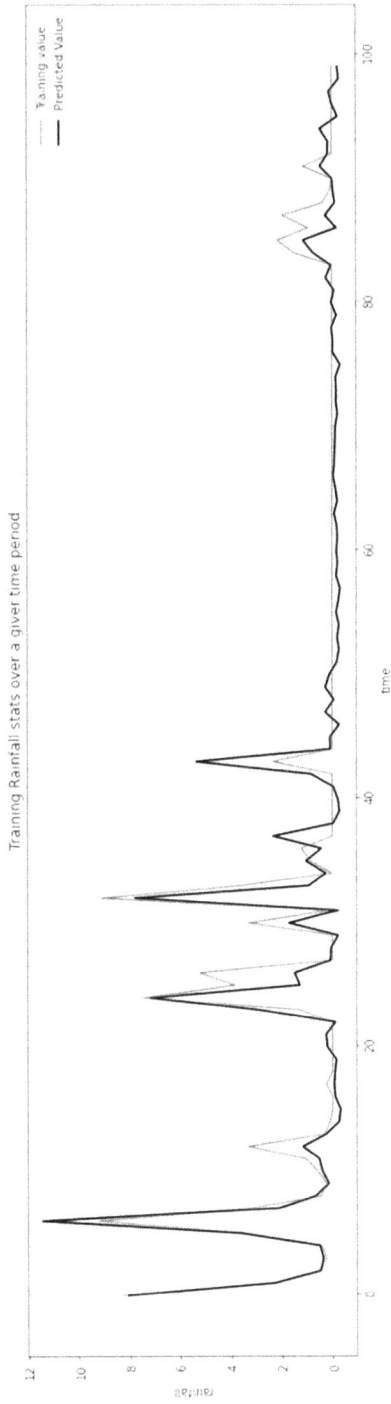

Figure 1.22 Actual vs. predicted values for training data.

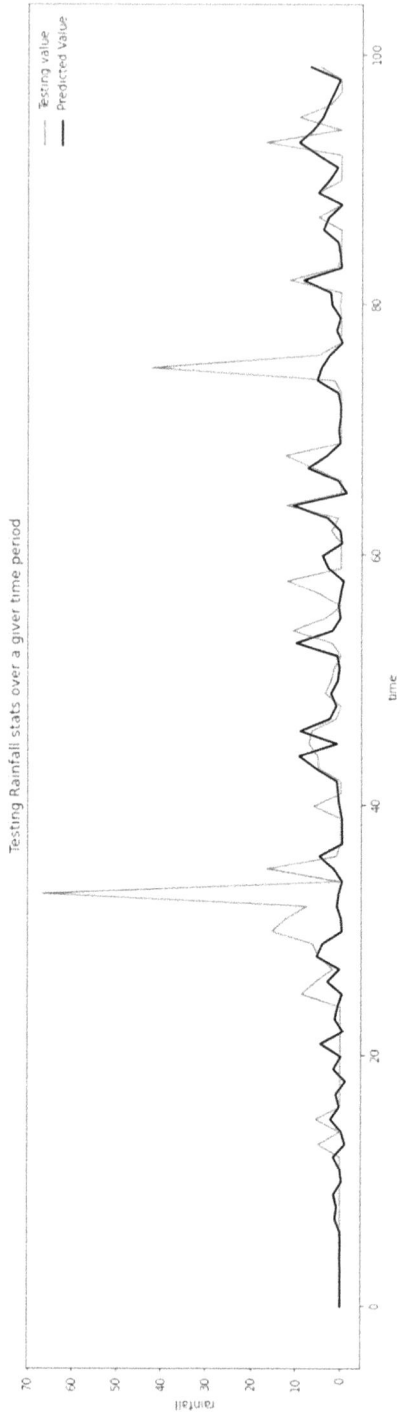

Figure 1.23 Actual vs. predicted values for testing data.

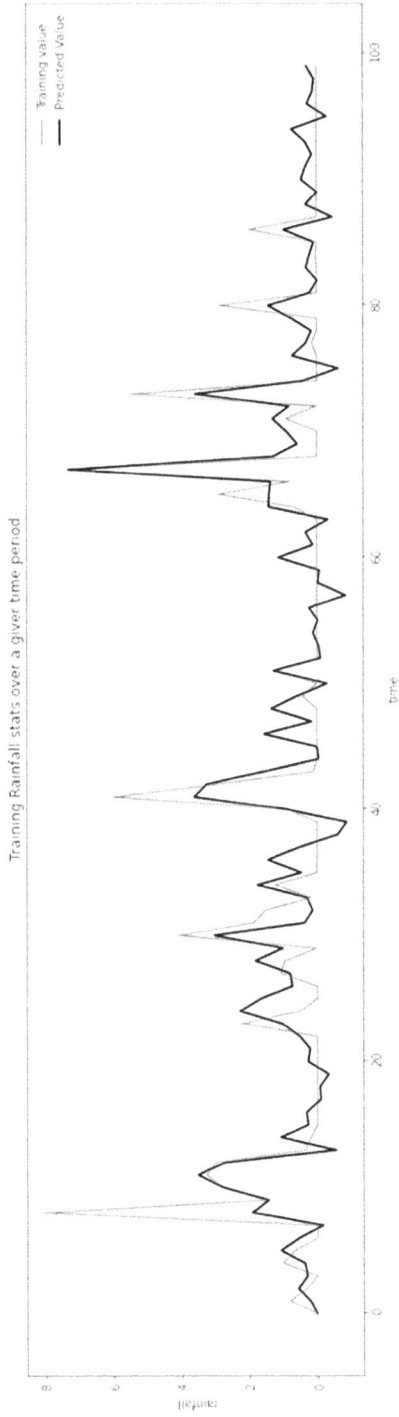

Figure 1.24 Actual vs. predicted values for training data.

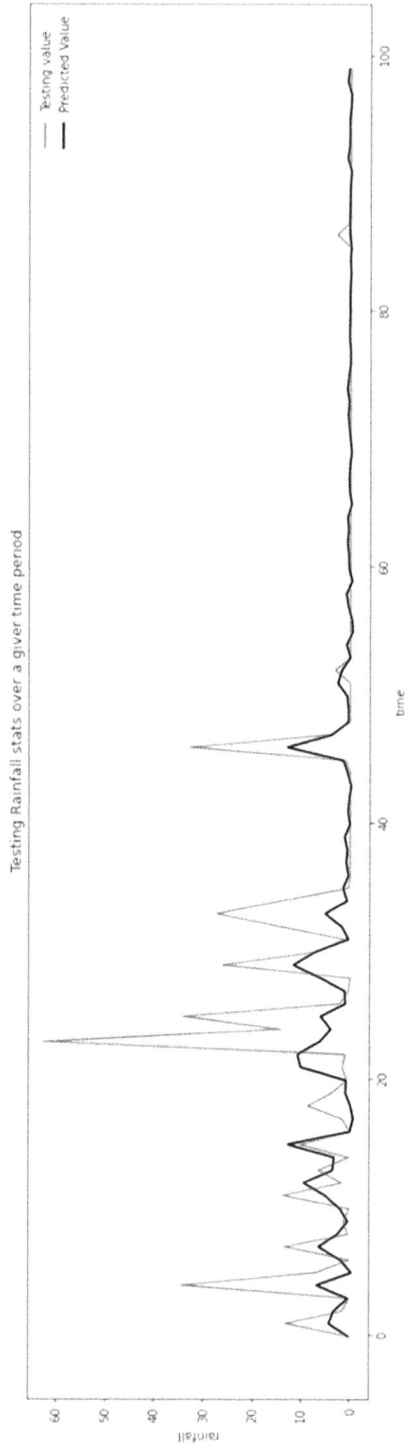

Testing Rainfall stats over a given time period

Figure 1.25 Actual vs. predicted values for testing data.

oversamples the observation in the given set and undersamples the remaining observation, thus giving a more balanced distribution. Time-Series GAN [14] generate new observations while preserving the original relationships between variables across time. The model not only captures the distribution of features at a point in time but also captures potential dynamics across time. TimeGAN consists of four features – embedding function, recovery function, sequence generator and sequence discriminator. The first two components are trained jointly with adversarial components (latter two) so that the model learns to encode features, generate representations and iterate across time.

BIBLIOGRAPHY

[1] Sepp Hochreiter and Jurgen Schmidhuber. "Long Short-Term Memory". *Neural Computation* 9.8 (Nov. 1997), pp. 1735–1780.

[2] SK Roy Bhowmik and VR Durai. "Application of multimodel ensemble techniques for real time district level rainfall forecasts in short range time scale over Indian region". *Meteorology and atmospheric physics* 106.1 (2010), pp. 19–35.

[3] AK Sahai, MK Soman, and V Satyan. "All India summer monsoon rainfall prediction using an artificial neural network". *Climate dynamics* 16.4 (2000), pp. 291–302.

[4] Sulagna Gope et al. "Early prediction of extreme rainfall events: a deep learning approach". In: *Industrial Conference on Data Mining*. Springer. 2016, pp. 154–167.

[5] Kandula V Subrahmanyam et al. "Prediction of heavy rainfall days over a peninsular Indian station using the machine learning algorithms". *Journal of Earth System Science* 130.4 (2021), pp. 1–9.

[6] Meng-Hua Yen et al. "Application of the deep learning for the prediction of rainfall in Southern Taiwan". *Scientific reports* 9.1 (2019), pp. 1–9.

[7] Philipp Hess and Niklas Boers. "Deep learning for improving numerical weather prediction of heavy rainfall". *Journal of Advances in Modeling Earth Systems* (2022), e2021MS002765.

[8] Suman Ravuri et al. "Skilful precipitation nowcasting using deep generative models of radar". *Nature* 597.7878 (2021), pp. 672–677.

[9] Bhupendra Nath Goswami et al. "Increasing trend of extreme rain events over India in a warming environment". *Science* 314.5804 (2006), pp. 1442–1445.

[10] J. N. Liu, Y. Hu, J. J. You, & P. W. Chan "Deep neural network based feature representation for weather forecasting". In: *Proceedings of the International Conference on Artificial Intelligence (ICAI)*. The Steering Committee of The World Congress in Computer Science, Computer Engineering and Applied Computing (WorldComp), p. 1, 2014.

[11] Alex Graves. "Generating sequences with recurrent neural networks". *arXiv preprint arXiv:1308.0850* (2013).

[12] Sergey Ioffe and Christian Szegedy. "Batch normalization: Accelerating deep network training by reducing internal covariate shift". In: *International conference on machine learning*. PMLR. 2015, pp. 448–456.

[13] Nitesh V Chawla et al. "SMOTE: synthetic minority over-sampling technique". *Journal of artificial intelligence research* 16 (2002), pp. 321–357.

[14] Jinsung Yoon, Daniel Jarrett, and Mihaela Van der Schaar. "Time-series generative adversarial networks". *Advances in Neural Information Processing Systems* 32 (2019).

Chapter 2

Diagnostic model for wheat leaf rust disease using image segmentation

Deepak Kumar and Vinay Kukreja

Chitkara University Institute of Engineering and Technology,
Chitkara University, Punjab, India

CONTENTS

2.1 INTRODUCTION

Wheat is one of the most important crops in the world, producing a huge amount of grain for industry and manufacturing. If the wheat plant is damaged or suffers from infection or fungal diseases, the wheat yield quality is reduced, and this in turn reduces grain quality. Crop diseases contribute both directly and indirectly to loss of yield quality [1, 2].

DOI: 10.1201/9781003415466-2

Plant diseases in a crop are identified at the stem, stripe, and leaf levels. The exact location of the disease can be identified by an experienced psycho-pathologist or by using computer-assisted image segmentation techniques [3–5]. Traditional image processing loses contextual information about pre-diction during processing. Errors such as image measurement, image dis-play, and feature release [6, 7] have been found during image recognition or when multiple objects are found. Different segmentation algorithms are required for partitioning and for image separation [8, 9] in the case of dis-eased plants.

2.1.1 Image segmentation

During image processing, the image is divided into different segments based on their properties [10, 11]. Segmentation is the process of defining labels for each pixel in an image [5]. Image segments are partitioned according to their pixel intensity value [4].

2.1.2 Properties of segmentation

Segmentation algorithms are categorized into two types based on basic fea-tures of gray values:

1) **Discontinuity**: split points, lines, and edges of the image.
2) **Similarity**: thresholding, regional growth, region separation, and integration.

2.1.3 Image segmentation process

Following image enhancement, color, shape and textural features are extracted, and discontinuity and similarity features are found in the com-puter-assisted segmentation process. Images are segmented in accordance with two rules: (1) finding the same things in the image [12]; and (2) sepa-ration to find the continuous values of the pixel intensity in the image. The segmentation process is shown in Figure 2.1.

2.1.4 Object detection, classification, and segmentation

Object-detection algorithms find the target object location through real-time images [4]. Finding multiple objects with the same color in an image is known as semantic segmentation [13]. Finding objects with different colors is known as instance segmentation. Semantic segmentation is used to divide segments of different objects, and instance segmentation is used to classify

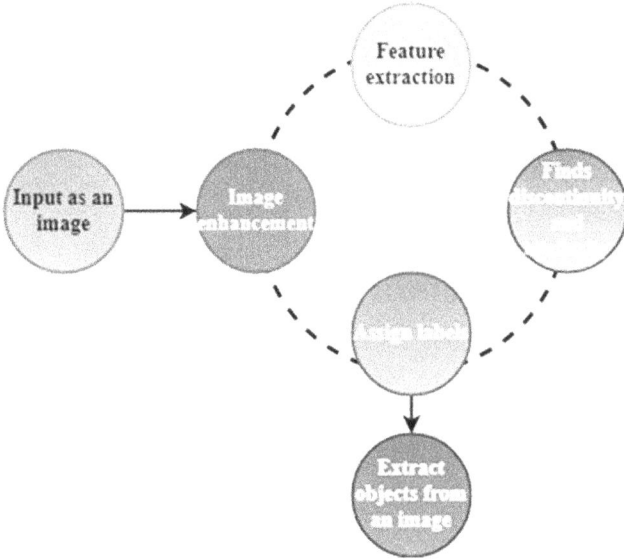

Figure 2.1 Segmentation process.

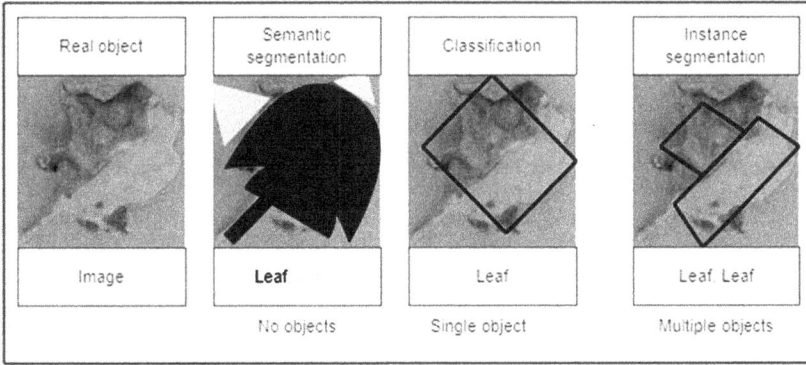

Figure 2.2 Comparison of semantic segmentation, classification, and instance segmentation in a leaf.

different objects in the same image. Instance segmentation therefore combines object detection and classification (Figure 2.2).

2.1.4.1 Semantic and instance segmentation

Semantic segmentation associates each pixel of the image with a defining class label, such as 'person'. It considers multiple objects of the same class

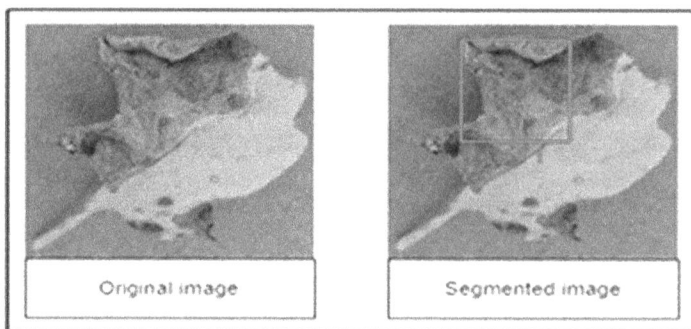

Figure 2.3 Example of instance segmentation.

as a single entity and defines a single entity with the same color. The same color entity therefore loses both contextual and global information about the predicted object. Instance segmentation, however, treats multiple objects of the same class as distinct individual instances [13], thus reducing information loss (Figure 2.3).

2.2 RELATED WORK

Many authors have used different types of segmentation techniques for detection or classification purposes. Semantic segmentation techniques such as K-means clustering and color threshold have been used to detect maize injury [3, 14]. Once the infection is transmitted, the whole plant is damaged. Features of the fusarium head blight (FHB) wheat disease [15, 16] have been recognized through K-means clustering and random-forest machine-learning algorithms. The infected regions of rice [7, 8] have been found using K-means clustering algorithm; infected areas of rice and groundnut [17] were found through multi-level Otsu thresholding and K-means clustering algorithm. Multi-level Otsu thresholding separates all regions within its pixel value and combines all the pixels that define the same infection regions. Powdery mildew [1] wheat disease has been identified through a local thresholding algorithm that converts healthy and diseased regions into gray-level pixels which are identified through mean pixels value. Wheat rust urediniospores [2] were found using the remote image-sensing watershed segmentation technique. Authors in [18, 19] found FHB-infected regions in wheat spikes using Mask convolutional neural networks (Mask-RCNN): each infected region and its boundary are identified in the form of color notation.

Table 2.1 shows the various segmentation techniques and tasks that have been performed by different researchers on different crops.

Table 2.1 Summary of various segmentation techniques applied on different crops

Reference	Year	Techniques	Planting crop	Images	Description
[14]	2020	K-means clustering, Color threshold application	Maize	30 RGB	The authors find the injury ratio in maize leaves through K-means clustering and threshold segmentation technique with different color spaces. During injury ratio identification, color threshold segmentation is more efficient than K-means clustering in HSV color space.
[3]	2018	Threshold	Maize	106 RGB	The infected rust spots of maize leaves are found through threshold techniques.
[15]	2020	K-means clustering, Random forest classifier	Wheat	6500 RGB	The author finds the severity of fusarium head blight wheat disease through K-means clustering and random forest classifier. Spots of different severity on wheat ears are counted using a random classifier that achieves 92.64% counting accuracy.
[7]	2021	Threshold	Rice	366 RGB	Local thresholding techniques applied to find infected regions on rice leaf images.
[16]	2019	K-means clustering	Wheat	1720 RGB	Fusarium head blight infected region in wheat is easily found with the help of K-means clustering.
[8]	2017	K-means clustering	Rice	50 RGB	Infected areas in healthy rice plants are found through a k-means clustering algorithm applied to 50 plant RGB images.
[17]	2021	Multi-level Otsu, K-means clustering	Rice, Groundnut, Apple	15 RGB images of each crop	The authors find affected areas of rice, groundnut, and apple from a complex background through multi-level Otsu and K-means clustering algorithms. Multi-level Otsu thresholding technique achieves better segmentation results than K-means clustering.
[9]	2018	K-means clustering with graph cut	Rice	25 UAV	The authors proposed the k-means clustering with graph cut algorithm for estimation of rice yield area. The k-means clustering algorithm is used to segment the rice grain areas. The graph cut algorithm is used to extract the grain area information

(Continued)

Table 2.1 (Continued) Summary of various segmentation techniques applied on different crops

Reference	Year	Techniques	Planting crop	Images	Description
[1]	2020	Threshold	Wheat	75 RGB	The authors identify the infected area of powdery mildew disease in wheat plant using a local threshold technique, achieving 93.33% segmentation accuracy.
[2]	2018	K-means clustering, Watershed	Wheat	120 UAV	Urediniospores in stripe rust are counted using K-means clustering and watershed techniques. Accuracy of 92.6% is achieved.
[10]	2017	Threshold	Wheat	20 RGB	The researchers recognize powdery mildew disease in wheat through the threshold segmentation technique.
[11]	2021	Threshold	Maize	19 UAV	Two threshold segmentation techniques, misclassification probability and the intersection method, are applied to maize under different water stress levels.
[20]	2018	K-means clustering	Wheat	360 RGB	With the help of the K-means clustering technique, the infected area of powdery mildew, tan spot, pink snow mold, and septoria leaf spot is calculated. Thus, the K-means clustering algorithm helps find the infected area of disease in the wheat plant.
[4]	2020	Montecarlo-sampled K-means segmentation	Rice	12 NIR	The ground mass evaluation in rice crops is found through the Montecarlo-sampled K-means segmentation algorithm using NIR images in different rice development stages.
[5]	2019	Color, column	Rice	72 Plots	The biomass of rice crops is estimated through color and column segmentation techniques, which help to identify the relationship of vegetation indices.

Ref	Year	Technique	Crop	Dataset	Description
[21]	2021	Threshold, Fuzzy means clustering	Maize	127 RGB	The diseased leaf area was extracted with the help of the threshold segmentation technique. The leaf area index helps find the severity using fuzzy decision rules.
[12]	2018	Threshold improved Otsu	Rice	898 RGB	Paper discusses rice seedling segmentation for paddy fields. Rice seedlings are extracted through Otsu segmentation. The RGB images are converted to YCRCB color space.
[13]	2021	Otsu multi-level thresholding, Fuzzy c-means, Fast k-means, Multi-level thresholding	Rice	24 Grains	The authors compare four segmentation techniques for the estimation of chalkiness in brown rice grains. The fuzzy c-means achieves higher accuracy than other segmentation techniques.
[22]	2020	Otsu, Gray level co-occurrence matrices, K-nearest neighbors	Rice	190 RGB	Rice leaf diseases are identified through Otsu and k-nearest neighboring algorithms. First, the Otsu segmentation technique is used to segment the regions. After extraction of regions, the color features are extracted through GLCM. After feature extraction, the uproot features have been predicted through the K-nearest neighbor algorithm.
[18]	2020	Mask-RCNN	Wheat	922 RGB	Fusarium head blight disease in wheat spikes is identified through the Mask-RCNN technique.
[19]	2021	Mask-RCNN	Wheat	524 RGB	Wheat spike segmentation and identification of FHB disease is carried out using Mask-RCNN technique, achieving 77.81% accuracy.

2.3 MATERIALS AND METHODS

This section introduces a proposed approach to segmentation, together with performance analysis parameters, and the various types of dataset available.

2.3.1 Dataset

A dataset is very important for locating symptoms of disease in plants, and is applied as an input to any segmentation model. There are two types of dataset: publicly available and private. A total of 3432 images have been taken from the global wheat detection dataset which has been used for wheat leaf rust identification purposes.

2.3.2 Segmentation techniques

There are four types of segmentation technique (Figure 2.4).

2.3.2.1 Pixel-based segmentation

The main purpose of pixel-based segmentation is to divide the image pixels based on their intensity value [22].

Thresholding segmentation. Thresholding is a simple segmentation technique that is used to create a binary image. The technique can be local, global or adaptive threshold [13]. The global threshold depends entirely on the image histogram. Image histograms may be affected by sound, contrast, color, space-filling, shadow, etc. The global boundary is determined using local image formats. Local property values can be used to improve the

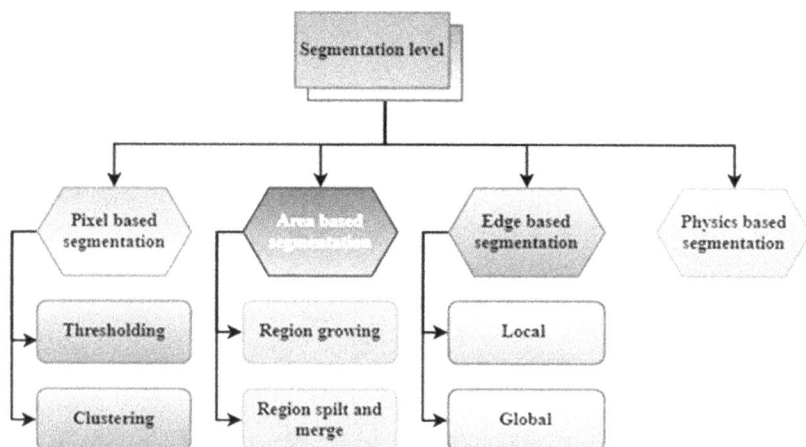

Figure 2.4 Segmentation-level techniques.

histogram and to calculate the exact boundary of the image [12]. Global threshold selection strategies based on local architecture are divided into two categories: histogram development methods and threshold computing methods. Image histograms can be enhanced so that the process of selecting the threshold is simplified, while limited computer systems seek to integrate the image area to calculate the total threshold value [4]. Local thresholding converts the black pixels of an image to white pixels. Otsu thresholding is an adaptive segmentation technique.

Clustering-based segmentation techniques. Clustering is the process of grouping objects that are similar to other objects into the same circle (called a collection) [21]. Clustering-based segmentation techniques can be K-means clustering or fuzzy c-means clustering.

2.3.2.2 Area-based segmentation

Area-based segmentation works on the regions or boundaries of an image and may involve the region growing or the region split-and-merge technique. Location-based classification looks for similarities between neighboring pixels and groups pixels with similar properties into different regions [11]. Areas are enlarged by combining pixels depending on their characteristics, such as thickness, varying below a specified value.

2.3.2.3 Edge-based segmentation

An edge filter is applied to the image, causing the pixels to be classified as either edge or edgeless. The edge can be detected based on the gray level, color, texture, brightness, saturation and contrast of the image. When edge discontinuities vary by gray level, color, or contrast, edge boundaries are marked [12].

2.3.2.4 Physics-based segmentation

This segmentation is applied when there are multiple types of objects in an image. The computer generates one or more descriptions of the event element that creates the image, especially the type of material and light [2, 18, 19]. It allows for the effective use of one or two that are easily separated to create a simple translation of human activity. This segmentation applies when one object in an image is dark in color and another is gray.

2.4 PROPOSED APPROACH

To overcome the issues of different segmentation-level techniques, the Mask-RCNN model is used to detect wheat leaf rust disease (see Figure 2.5).

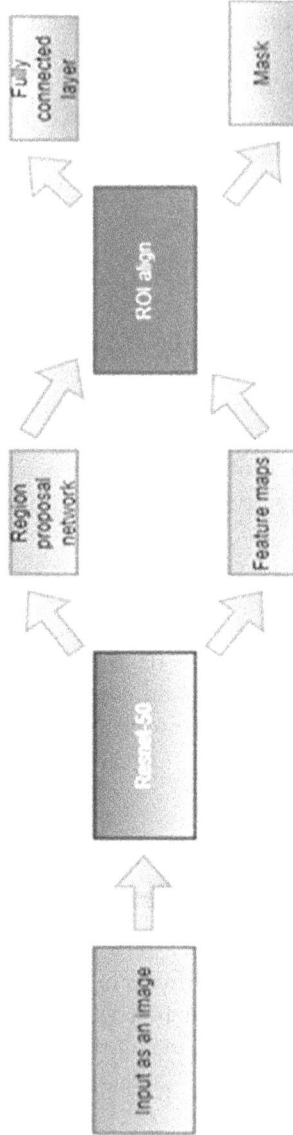

Figure 2.5 Overview of Mask-RCNN.

In the Mask-RCNN model, the input is taken as an image and the features are extracted using the Resnet-50 model [19]. After feature extraction, the region proposal network generates the mask of the binary class of each object through the regression and classification layer. The region of interest (ROI) generates different masks for each object. Once all the binary masks of each leaf have been generated, they are classified using Softmax and the bounding box location of each leaf is determined by regression.

2.5 PERFORMANCE PARAMETERS

Several parameters are used to evaluate each segmentation technique, including mean average precision (mAP), mean average recall (mAR) and F1-score.

2.6 EXPERIMENTAL SETUP

All the segmentation experiments were performed on a Sony Vaio laptop with a Core i5 processor with Gforce. For experimentation purposes, a total of 4500 images were taken from the Kaggle website. The dataset used is publicly available. Table 2.2 summarizes the materials used in this study. Table 2.3 shows the parameters used by Mask-RCNN.

Table 2.2 Summary of materials used in this study

Hardware	Software	Dataset
Core i5 processor Nvidia Gforce	Python 3.6.1 Matplotlib library cv2 library Jupyter notebook	Global wheat detection (https://www.kaggle.com/c/global-wheat-detection)

Table 2.3 Parameters of Mask-RCNN

Parameters	Value
Backbone	Resnet50
Image Min_dimension	1024 × 1024
Image Max_dimension	256 × 256
Scale of anchor	[8, 16, 32, 64, 128]
The aspect ratio of the anchor	[0.5, 1]
RPN threshold	0.8

2.7 EXPERIMENTAL RESULTS

All the segmentation experiments were performed using Python and implemented on a Jupyter notebook. Python has high speed for pixel computation and includes libraries such as scipy, Matplotlib, and opencv which were used to implement the Mask-RCNN model. A total of 3000 images were used for leaf rust detection using various segmentation techniques. Firstly, the RGB image is taken as an input to threshold and Otsu thresholding segmentation techniques, which convert it into a gray-scale image, transforming the pixels into different gray and white colors. The threshold segmentation technique calculates the threshold value of each gray pixel and estimates the effective results of leaf rust on the wheat leaf. A fuzzy c-means clustering algorithm is used to convert the image into gray scale and finds the casual relationships of each pixel in each center. Once the center of an image is found, it makes a cluster of similar centers. Fuzzy c-means clustering is a better algorithm than K-means. In the K-means algorithm, the data points belong exclusively to one cluster, whereas in the c-means fuzzy algorithm, a data point can belong to more than one cluster with a certain probability. Fuzzy c-means creates k numbers of clusters and then assigns each data point to a cluster, but it will be a factor that will define how strongly the data belongs to that cluster. The experimental results of segmentation-level techniques are shown in Figure 2.6.

The images were taken from the global wheat detection dataset, freely available on Kaggle. One-tenth of the 3432 images were randomly selected for training purposes. The training images have a strong effect on prediction. Several hyperparameters were used in Mask-RCNN during training, and a minimum value of 0.5 was set for each ROI. Any ROI with a threshold value less than 0.5 was rejected and not used for training purpose. The standard size of images was set to be 256 × 256 pixels. If images are of different dimensions, the training accuracy of Mask-RCNN is reduced due to GPU image speed computation, with increased consequential losses. Two images are passed to the GPU at the same time, meaning that 20,000 iterations have been used for GPU training for the estimation of different leaf locations on the wheat plant. Once the leaf location is determined, it is easy to recognize rust disease. The mask branch is applied to the 100 highest-scoring detection boxes, speeding up inference and accuracy.

The red color shows the whole wheat leaf, while the location of each leaf rust is denoted by blue color. The average precision and average recall on wheat leaf has been denoted on each leaf rust boundary location. The segmentation results are shown in Figure 2.7.

During recognition of wheat leaf rust, three different types of losses occurred: MRCNN classification, MRCNN box, and MRCNN Mask. These three losses coincide with the MRCNN baseline on the global wheat

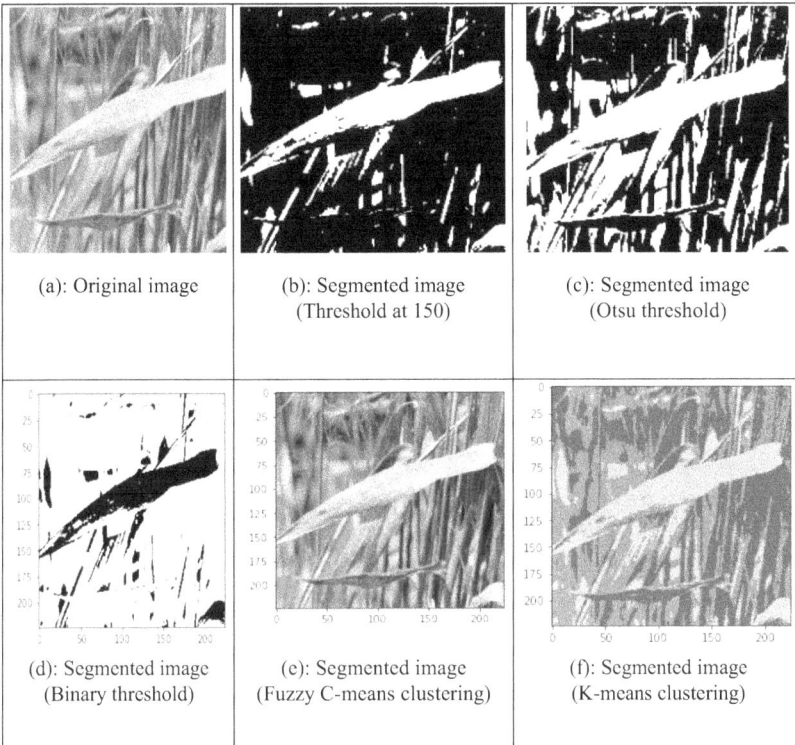

(a): Original image

(b): Segmented image
(Threshold at 150)

(c): Segmented image
(Otsu threshold)

(d): Segmented image
(Binary threshold)

(e): Segmented image
(Fuzzy C-means clustering)

(f): Segmented image
(K-means clustering)

Figure 2.6 Experimental results.

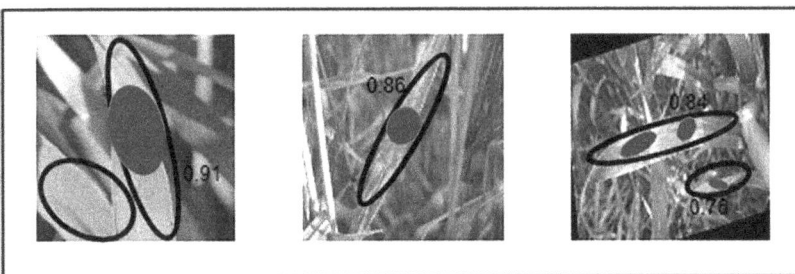

Figure 2.7 Segmentation results of Mask-RCNN where red color shows the ground truth label and yellow color shows segmentation results for wheat leaf rust.

Table 2.4 Comparison of Otsu, k-means, fuzzy c-means, and our proposed model in terms of their accuracy

Model	mAP	mAR	F1-score
Otsu	0.56	0.82	0.52
K-means	0.49	0.832	0.56
Fuzzy c-means	0.45	0.846	0.59
Proposed model (Mask-RCNN)	0.59	0.78	0.96

detection dataset. The loss for classification and box regression is the same as Faster R-CNN.

- A per-pixel sigmoid is applied to each map
- The map loss is then defined as the average binary cross-entropy loss
- Mask loss is only defined for the ground truth class
- Separates class prediction and mask generation
- Empirically better results and the model is easier to train

2.8 PERFORMANCE ANALYSIS

The performance of each segmentation technique as measured by mean average precision (mAP), mean average recall (mAR), and F1-score is shown in Table 2.4. Our proposed model, when compared with Otsu, k-means, and fuzzy c-means segmentation techniques, is shown to have high mAP, high mAR, and high F1-scores.

2.9 CONCLUSION

Diseases in wheat crops, such as powdery mildew, brown rust, spot disease, and snow mold, are increasing daily. According to the National Agriculture Research Institute, wheat quality loss, which in turn affects wheat grain quality and production rate, has been increasing at a rate of 6–7% annually. This paper has reviewed different types of segmentation techniques for maize, rice, and wheat crops. All segmentation techniques have been implemented on the global wheat detection dataset using Python. During Mask-RCNN model implementation, three types of losses occurred on the MRCNN baseline: boundary, class, and mask. Three different performance constants were used for evaluation of wheat leaf rust identification segmentation techniques: mAP, mAR, and F1-score. Our proposed model achieves a high F1-score (0.96) as compared to the fuzzy c-means clustering (0.59), K-means (0.56), and Otsu (0.52) segmentation techniques. The proposed work has been tested on wheat plant images and the quantitative results show good segmentation without manual intervention.

REFERENCES

[1] L. H. Jinling Zhao, Y. Fang, G. Chu, and H. Yan, "Identification of Leaf-Scale Wheat Powdery Mildew (Blumeria graminis f. sp. Tritici) Combining Hyperspectral Imaging and an SVM Classifier," *Plants*, vol. 9, no. 8, pp. 1–13, 2020.

[2] Y. Lei, Z. Yao, and D. He, "Automatic Detection and Counting of Urediniospores of Puccinia striiformis f. sp. Tritici Using Spore Traps and Image Processing," *Sci. Rep.*, vol. 8, no. 1, pp. 1–11, 2018.

[3] A. Yadav and M. K. Dutta, "An Automated Image Processing Method for Segmentation and Quantification of Rust Disease in Maize Leaves," in *International Conference on Computational Intelligence and Communication Technology*, 2018, pp. 1–5.

[4] J. D. Colorado et al., "A Novel NIR-Image Segmentation Method for the Precise Estimation of Above-Ground Biomass in Rice Crops," *PLoS One*, vol. 15, no. 10, pp. 1–20, 2020.

[5] C. A. Devia, J. P. Rojas, E. P. Carol, M. Ivan, F. M. D. Patino, and J. Colorado, "High-Throughput Biomass Estimation in Rice Crops Using UAV Multispectral Imagery," *J. Intell. Robot. Syst.*, vol. 96, no. 3–4, pp. 573–589, 2019.

[6] P. Wang, Y. Zhang, B. Jiang, and J. Hou, "An Maize Leaf Segmentation Algorithm Based on Image Repairing Technology," *Comput. Electron. Agric.*, vol. 172, pp. 105349–105362, 2020.

[7] A. Islam, R. Islam, S. M. R. Haque, S. M. M. Islam, and M. A. I. Khan, "Rice Leaf Disease Recognition using Local Threshold Based Segmentation and Deep CNN," *Int. J. Intell. Syst. Appl.*, vol. 13, no. 5, pp. 35–45, 2021.

[8] P. Kumar, B. Negi, and N. Bhoi, "Detection of Healthy and Defected Diseased Leaf of Rice Crop using K-Means Clustering Technique," *Int. J. Comput. Appl.*, vol. 157, no. 1, pp. 24–27, 2017.

[9] M. N. Reza, I. S. Na, S. W. Baek, and K. H. Lee, "Rice Yield Estimation based on K-Means Clustering with Graph-Cut Segmentation Using Low-Altitude UAV Images," *Biosyst. Eng.*, vol. 177, pp. 109–121, 2019.

[10] Z. Diao, C. Diao, and Y. Wu, "Algorithms of Wheat Disease Identification in Spraying Robot System," in *9th International Conference on Intelligent Human-Machine Systems and Cybernetics, IHMSC 2017*, 2017, vol. 2, pp. 316–319.

[11] Y. Niu, H. Zhang, W. Han, L. Zhang, and H. Chen, "A Fixed-Threshold Method for Estimating Fractional Vegetation Cover of Maize Under Different Levels of Water Stress," *Remote Sens.*, vol. 13, no. 5, pp. 1–19, 2021.

[12] J. Liao, Y. Wang, J. Yin, L. Liu, S. Zhang, and D. Zhu, "Segmentation of Rice Seedlings Using the YCrCB Color Space and an Improved Otsu Method," *Agronomy*, vol. 8, no. 11, pp. 1–16, 2018.

[13] W. Wongruen, "Comparative Study Estimating and Detecting Chalkiness of Thai Hom Mali Brown Rice Grains Using Image Analysis and Four Segmentation Techniques," in *18th International Conference on Electrical Engineering/Electronics, Computer, Telecommunications and Information Technology: Smart Electrical System and Technology*, 2021, pp. 392–399.

[14] A. H. Abbas, N. M. Mirza, S. A. Qassir, and L. H. Abbas, "Maize Leaf Images Segmentation Using Color Threshold and K-means Clustering Methods to Identify the Percentage of the Affected Areas," *IOP Conf. Ser.: Mater. Sci. Eng.*, 2020, vol. 745, no. 1, pp. 1–11.

[15] D. Zhang et al., "Evaluation of Efficacy of Fungicides for Control of Wheat Fusarium Head Blight Based on Digital Imaging," *IEEE Access*, vol. 8, pp. 109876–109890, 2020.

[16] W. Fusarium et al., "Using Neural Network to Identify the Severity of Wheat Fusarium Head Blight in the Field Environment," *Remote Sens.*, vol. 11, no. 20, pp. 2375–2392, 2019.

[17] B. R. Prasad et. al., "Performance Comparison of Unsupervised Segmentation Algorithms on Rice, Groundnut, and Apple Plant Leaf Images," *Inf. Technol. Ind.*, vol. 9, no. 2, pp. 1090–1105, 2021.

[18] W. Su et al., "Evaluation of Mask RCNN for Learning to Detect Fusarium Head Blight in Wheat Images," in *An ASABE Meeting Presentation*, 2020, pp. 1–3.

[19] W. H. Su, J. Zhang, C. Yang, R. Page, T. Szinyei, C. D. Hirsch, and B. J. Steffenson, "Automatic Evaluation of Wheat Resistance to Fusarium Head Blight Using Dual Mask-RCNN Deep Learning Frameworks in Computer Vision," *Remote Sens.*, vol. 13, no. 1, pp. 26–42, 2021.

[20] S. Nema and A. Dixit, "Wheat Leaf Detection and Prevention Using Support Vector Machine," in *International Conference on Circuits and Systems in Digital Enterprise Technology, ICCSDET 2018*, 2018, pp. 1–5.

[21] M. Sibiya and M. Sumbwanyambe, "Automatic Fuzzy Logic-Based Maize Common Rust Disease Severity Predictions with Thresholding and Deep Learning," *Pathogens*, vol. 10, no. 2, pp. 1–17, 2021.

[22] N. Manohar and K. J. Gowda, "Image Processing System based Identification and Classification of Leaf Disease: A Case Study on Paddy Leaf," in *Proceedings of the International Conference on Electronics and Sustainable Communication Systems*, 2020, pp. 451–457.

Chapter 3

A comparative study of traditional machine learning and deep learning approaches for plant leaf disease classification

R Ahila Priyadharshini, S Arivazhagan and M Arun
Mepco Schlenk Engineering College, Sivakasi, India

CONTENTS

3.1 INTRODUCTION

Plants are essential to our continued existence because they not only feed us but also protect us from the harmful effects of radiation. Life could not exist without plants; they provide food and shelter for all living beings on earth and protect the ozone layer, which blocks the harmful effects of ultraviolet radiation. Agricultural productivity is crucial to the global economy. Crop disease is a natural occurrence that must be promptly identified and treated with precision. There can be severe consequences for plants if this is not carried out at the appropriate time, affecting product quality, quantity, and productivity.

The majority of plant diseases manifest through their leaves. Pathologists diagnose plant diseases by performing an examination of the leaves of affected plants. Manually diagnosing a plant disease is an extremely time-consuming process, and the accuracy of a diagnosis is directly proportional to the skill of the plant pathologist performing the examination. This makes the classification of plant leaf diseases an ideal application for computer-aided diagnostic systems.

The primary focus of this chapter is on determining the most common diseases that can affect tomato leaves, because tomato is an edible vegetable widely consumed and grown all over the world. Consumers across the globe consume more than 160 million tons of tomatoes annually [1]. Tomatoes contain one of the highest concentrations of nutrients, so their production has a significant effect on the agricultural economy. Tomatoes are increasingly in demand; in addition to the high nutrient content, they are known to have pharmacological properties that are effective against a range of diseases, including gingival bleeding, gingivitis, and hypertension [2].

According to Stilwell, "small farmers are responsible for more than 80% of agricultural output; however, nearly 50% of the crops grown by these farmers are destroyed by pests and diseases" [1]. Diseases caused by parasitic insects are the primary causes affecting tomato growth, necessitating field crop disease diagnosis research. Five tomato diseases are caused by insects and 16 by bacteria, fungi, or poor agricultural practices. Bacterial wilt is a devastating disease caused by *Ralstonia solanacearum* bacteria. This has a long lifespan in soil and can enter roots through wounds caused by secondary root formation, cultivation, transplanting, or even insects.

The development of disease is also encouraged by conditions of high temperature and humidity. As bacteria spread rapidly throughout the plant's water-conducting tissue, the plant quickly became covered in a slimy substance, which affects its vascular system even though the leaves may still appear green. When viewed in cross-section, the stem of an infected plant looks brown and yellowish. Once we have determined which parts of the plant are infected with disease, we must search for changes such as black or brown patches and finally, we must look for insects.Several researchers have used machine learning and neural network architectures to identify plant diseases. The majority of modern techniques for plant leaf disease recognition rely on color [3–5], shape [6] and textural [7] characteristics. Common machine learning (ML) techniques for classifying plant diseases include neural networks [8, 9], logistic regression, random forest [10, 11], support vector machines [12], adaptive boosting (AdaBoost) [11], k-nearest neighbors (kNN), and Näive Bayes [12]. However, these conventional methods typically involve two steps and rely heavily on the provision of hand-crafted features; they are costly and time-consuming because the expert must laboriously extract these characteristics from the images based on shape, texture, color, and size. Basavaiah and Arlene Anthony described a novel strategy using the combination of multiple features by combining Hu moments, Haralick and color histograms with local binary pattern characteristics. The features derived were classified using decision trees and random forest algorithm, achieving 90% and 94% accuracy respectively [13]. Kalyoncu et al. [14] proposed a novel classification technique for plant leaves using the combination of multiple descriptors. The extracted descriptors are the properties of shape, texture, geometry, and color. In addition, sorted uniform LBP (a novel variation of LBP) was proposed for the description of leaf texture. After combining the retrieved characteristics, the linear discriminant classifier (LDC) was used as the classifier. The three datasets analyzed for this method were ICL, Flavia, and Swedish, with average accuracy of 86.8%, 98.6%, and 99.5% respectively. Kaur et al. suggested a semi-automatic classification technique for soybean leaf diseases that combines texture and color data with an SVM classifier and achieved 90% accuracy with 4775 images [15].

Although there are numerous applications for machine learning techniques, feature engineering remains the greatest obstacle. Deep neural networks (DNNs) have enabled the development of promising plant pathology solutions without the need for time-consuming hand-crafted feature extraction. DNNs greatly enhance the accuracy of image classification. In recent years, applications employing deep learning for computer vision have advanced significantly. Convolutional neural networks (CNNs) are widely employed for medical applications [16, 17], speech processing [18], character recognition [19], and plant disease classification in tomato, rice, cucumber and maize leaf, etc. Brahimi et al. suggested a deep model approach for categorizing leaf diseases in tomato plants. Their experiments demonstrated the value of using a pre-trained model, particularly when there are fewer training samples, as opposed to the context of disease classification. They also recommended using occlusion techniques to localize the disease regions, aiding manual comprehension of the disease [20]. Trivadi et al. proposed a DNN model to detect and categorize tomato plant leaf diseases and provided a detailed discussion of biotic diseases caused by bacterial and fungal pathogens, including "blight, blast, and browns of tomato leaves" [21]. Ouhami et al. [22] considered the transfer learning capabilities of deep learning models such as DenseNet161, DenseNet121 and VGG16 for classifying plant leaves infected by six distinct types of pest and plant diseases, and achieved accuracy of 95.65%, 94.93%, and 90% respectively. Yang et al. presented a novel technique for identifying diseases on paddy leaves based on DNNs. CNNs were trained to identify 10 common rice diseases using 500 images of healthy and diseased paddy leaves and stems taken in an experimental paddy field. Using a 10-fold cross-validation strategy, the proposed CNN model achieved an accuracy of 95.48% [23].

Kawaski et al. introduced a unique system based on CNNs for detecting plant diseases. The authors used a total of 800 cucumber leaf images to train the CNN. The proposed CNN-based system classified cucumbers into three classes (two disease classes and one non-disease class) with an average accuracy of 94.9% [24] using four-fold cross-validation. Ahila Priyadharshini et al. proposed a modified Le-Net for maize leaf disease classification, using images from the Plant Village dataset for their experiments. CNNs are trained to differentiate between four distinct classes (three disease classes and one healthy class). The model's accuracy was 97.89% [25]. Zaki et al. [26] used a refined CNN model based on transfer learning and MobileNetV2 to classify tomato leaf diseases. The MobileNetV2 model accurately identified the disease more than 90% of the time. Agarwal et al. [27] proposed a novel CNN model with an average accuracy of 91.2% for classifying tomato diseases. A CNN model with Inception modules and dilated convolution using the PlantVillage dataset was proposed, achieving accuracy of 99.37% [28]. The authors of [29] proposed a lightweight CNN model with eight hidden layers from tomato plant leaf images from the PlantVillage dataset, which recognized diseases with 98.4% accuracy. The authors of [30] created a nine-layer custom-designed CNN model for recognizing 39 types of plant leaves from the PlantVillage dataset. Several augmentation techniques were used to increase the number of training images, and the custom-designed CNN obtained an accuracy of 96.46%. Nithish et al. [31] proposed a pre-trained ResNet-50 model as a transfer learning technique for disease detection in tomato leaf. After pre-training, ResNet-50 was modified to classify six distinct categories with a 97% accuracy rate.

3.2 MATERIALS AND METHODS

3.2.1 Dataset used

For this study, we used the tomato leaves in the PlantVillage dataset [32]. Although many diseases and pests can affect tomato plants, fungal diseases are very common. In this PlantVillage dataset, a total of ten classes of tomato leaves are available, nine diseased and one healthy (see Table 3.1). The total number of tomato leaf images across the 10 classes is 16011; 12005 are used for training and 4006 for testing, maintaining a 75:25 train/test ratio (Table 3.2). All the images in the dataset are of size 256×256.

3.2.2 Log-Gabor transform

The Gabor filter is a linear filter whose impulse response is a harmonic function modified by a Gaussian function. According to the uncertainty principle, it is optimally localized in both the spatial and frequency domains. However, it has two major disadvantages. Gabor filters have a maximum bandwidth of about one octave; consequently, they are not the best option if you're looking for broad-spectrum information with the highest degree of spatial localization. The Gabor transform would over-emphasize low-frequency components while under-representing high-frequency components [33].

As an alternative to the Gabor function, Field [34] presented the Log-Gabor function. The bandwidth of Log-Gabor filters is adjusted to make them as small as possible. Field asserts that filters with Gaussian transfer functions are more effective at concealing natural images when viewed on a logarithmic frequency scale. Two-dimensional Log-Gabor filters contain two frequency domain components: the radial component $G(f)$ which regulates the filter's response bandwidth, and the angular component $G(\varphi)$ which

Table 3.1 Details of tomato leaf diseases

S. no.	Tomato leaf	Sample leaf image	Description
1.	Target Spot		The disease begins on the elder leaves and spreads upward. Initial indicators are yellow-edged, irregularly shaped dots. The disease rapidly extents to all leaflets and other leaves, causing them to turn yellow and die. There are also spots on the stems.
2.	Mosaic Virus		Mosaic virus contaminates a wide variety of plants, but it is most prevalent in tomatoes. While the mosaic virus does not destroy the plant, it reduces both quantity and quality of the tomato fruits produced. The virus receives its name from the mosaic of light green and yellow marks on infected plants' leaves and mottling on their fruits. Leaves can also be shaped like ferns and grow in odd shapes.
3.	Curl Virus		The most destructive tomato disease is a DNA virus from the family *Geminiviridae* and the genus *Begomovirus*
4.	Bacterial Spot		Bacterial spot can infect all parts of the tomato plant, including leaf, stem, and fruit, except the roots. It manifests on leaves as water-soaked, microscopic, circular spots. Spots may initially appear yellow-green, but as they age, they turn brownish-red. When the disease is severe, there may be widespread leaf yellowing and leaf loss.
5.	Early Blight		Alternaria is a fungus that causes early blight in tomato plants. On the lower leaves, brown or black spots with dark edges appear, resembling a target. Fruit stem ends are susceptible to attack, resulting in the development of large, indented black spots with concentric rings. Typically, this fungus attacks plants after they have produced fruit.

(Continued)

Table 3.1 (Continued) Details of tomato leaf diseases

S. no.	Tomato leaf	Sample leaf image	Description
6.	Healthy		A healthy tomato leaf is softly fuzzed, medium-green.
7.	Late Blight		The fungus *Phytophthora infestans* causes the fast-spreading tomato plant disease late blight, which arises during cool, rainy conditions near the conclusion of the growth season. It appears on foliage as frost damage, with irregular green-black splotches. Large, irregular-shaped dark patches on fruits can swiftly destroy them.
8.	Leaf Mold		The fungus *Passalora fulva* causes leaf mold and is more common in greenhouse and high-tunnel tomatoes. High relative humidity is the cause of the sickness. Infected leaves wither and die, reducing yield indirectly. In severe cases, blooms and fruit may get diseased.
9.	Septoria Leaf Spot		Septoria leaf spot appears as small, circular spots with a grayish-white center and black margins. In the center of each imperfection, little black dots may appear. Leaves that have been affected turn yellow, wither, and fall off.
10.	Spider Mite		Spider mite damage to tomato leaves appears as a scattering of pale yellow spots on the leaf's upper surface. Eventually, the leaves will turn brown and die or fall off. The plant forms webs as a result of a severe attack.

controls the filter's orientation selectivity. $G(f)$ and $G(\varphi)$ are spelt out in Eqn. (3.1).

$$G(f,\varphi) = G(f) \times G(\varphi) = e^{-\frac{\left(\log\left(\frac{f}{f_0}\right)\right)^2}{2\left(\log\left(\frac{\sigma_f}{f_0}\right)\right)^2}} \times e^{-\frac{(\varphi-\varphi_0)^2}{2\sigma_\varphi^2}} \tag{3.1}$$

Table 3.2 Details of the tomato leaves used for the experimentation

S. no	Tomato leaf disease	# Images	# Training images	# Testing images
1.	Target Spot (TS)	1404	1053	351
2.	Mosaic Virus (MV)	373	279	94
3.	Curl Virus (CV)	3208	2405	803
4.	Bacterial Spot (BS)	2127	1595	532
5.	Early Blight (EB)	1000	750	250
6.	Healthy (H)	1591	1193	398
7.	Late Blight (LB)	1909	1431	478
8.	Leaf Mold (LM)	952	714	238
9.	Septoria Leaf Spot (SLS)	1771	1328	443
10.	Spider Mite (SM)	1676	1257	419

where f_0 is the center frequency of the filter, σ_f is the scaling factor of the filter's bandwidth, φ_0 is the orientation angle and σ_φ represents the angular bandwidth.

Log-Gabor functions have the following two distinguishing characteristics.

- They do not have a DC component (contrast of image ridges and edges is increased).
- They have a long tail at high frequencies so that broad-spectrum information with localized spatial breadth can be obtained (which helps preserve ridge structures).

Figure 3.1 demonstrates the result of convolving a sample leaf image with Log-Gabor filters with four scales and eight orientations to produce 32 sub-bands.

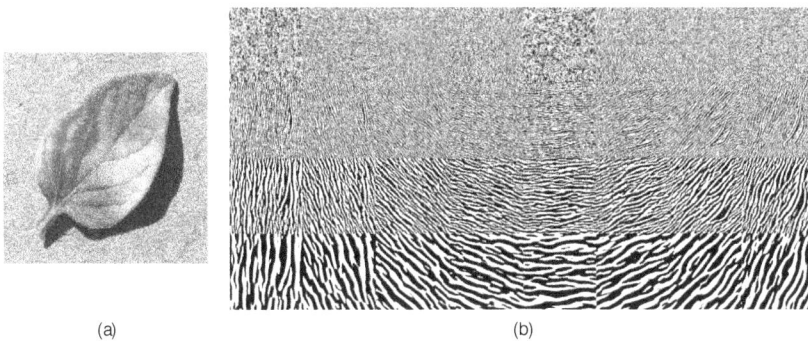

(a)   (b)

Figure 3.1 (a) Sample diseased tomato leaf image. (b) Log-Gabor transformed image with 4 scales and 8 orientations.

3.2.3 Convolutional neural networks

The ability of CNNs to recognize image patterns is well known. CNN has many convolutional and optional fully connected (FC) layers. CNNs employ filters to identify the characteristics of images. A filter is a matrix of weight values that has been trained to recognize precise features, such as corners, edges, etc. The filter applies the convolution operation to the image. The filter convolves with a portion of an image and will produce a high value if the feature exists in that portion of the image; otherwise, the value will be low. Using a stride value, the filter moves across the input image at variable intervals. The stride value indicates the step movement of the filter. The output image's dimension after the convolution is determined using Eqn. 3.2.

$$y = \text{floor}\left(\frac{x - w}{s}\right) + 1 \tag{3.2}$$

where x and y denote the size of input and output image respectively, w denotes filter size and s denotes stride.

The filter must be mapped non-linearly in order for CNN to understand the values of the filter. A nonlinear activation function processes the output of the convolution operation after adding it to a bias term. Networks can become nonlinear by using activation functions. We used the rectified linear unit (ReLU) activation function, as shown in Eqn. 3.3, because the tomato leaf input data was nonlinear. The ReLU activation function makes all negatives zero, leaving the positives as such.

$$f(X) = \max(0, X) \tag{3.3}$$

After a few layers of convolutional operation, CNNs use down-sampling to reduce the convolutional layer's output representation size. This accelerates the training process and reduces the network's memory consumption. There are numerous down-sampling methods, max pooling being the most prevalent. During max pooling, a window traverses an image based on the stride value and the output is the largest value within the window.

Each convolutional layer can optionally be followed by a batch normalizing layer in a CNN. Each layer's inputs are normalized by batch normalization, which reduces the internal coverable shift issue. In addition, to improve the stability of the network, it normalizes the output of a prior activation layer by subtracting the mean of the mini-batch and dividing it by the standard deviation of the same. Several convolutional layers and down-sampling

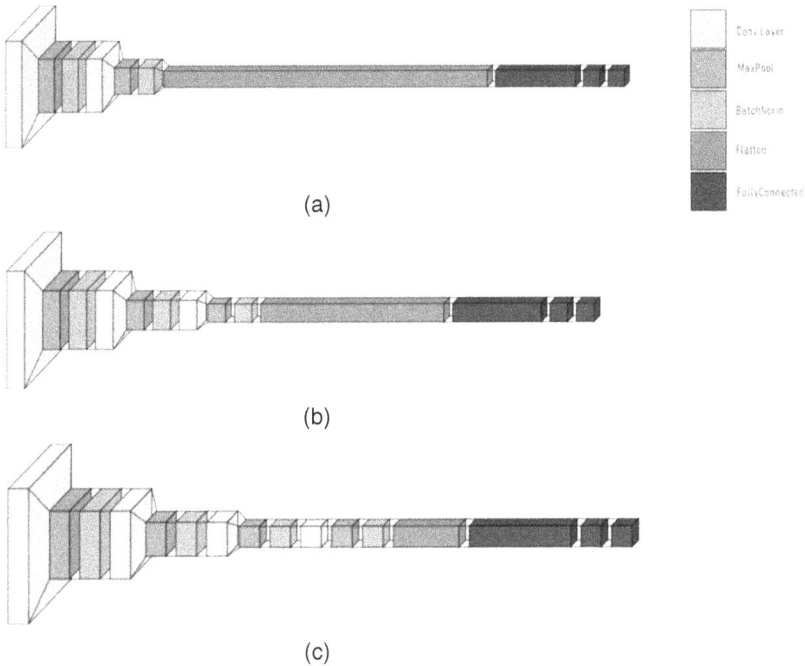

Figure 3.2 Custom CNN architecture (a) with depth 2 (b) with depth 3 (c) with depth 4.

operations are used to transform the image representation into a feature vector, and this is sent to the FC layers of a multi-layer perceptron. The output layer of a CNN produces the probability of the classes being predicted. To achieve this, the final layer will contain the same number of neurons as the number of classes.

Three different custom-designed CNNs are used in this chapter and the effect of the network's depth is studied (Figure 3.2). The feature map size and the number of learnable parameters for each layer are described in Table 3.3. The first 64 feature maps obtained in all four convolutional layers for a sample diseased leaf image are shown in Figure 3.3.

As we move deeper (left to right), this architecture reveals that the height and width tend to decrease, while the number of channels increases. All convolutional layers in the architecture employ 3 × 3 filters and ReLU activation. The primary advantage of using ReLU is that it does not activate all neurons simultaneously, thereby making the network sparse, efficient, and computationally simple. The Softmax activation function is used in the output layer because it is a general logistic activation function that can be used

Table 3.3 Details of the feature map size and the number of learnable parameters of 3 custom-designed CNNs

Custom CNN architecture

Layers	Depth 2		Depth 3		Depth 4	
	Output dimensions	# Learnable parameters	Output dimensions	# Learnable parameters	Output dimensions	# Learnable parameters
Input	32 × 32 × 3	—	32 × 32 × 3	—	32 × 32 × 3	—
Conv1	32 × 32 × 64	1792	32 × 32 × 64	1792	32 × 32 × 64	1792
MaxPool1	16 × 16 × 64	0	16 × 16 × 64	0	16 × 16 × 64	0
BatchNorm1	16 × 16 × 64	256	16 × 16 × 64	256	16 × 16 × 64	256
Conv2	16 × 16 × 96	55392	16 × 16 × 96	55392	16 × 16 × 96	55392
MaxPool2	8 × 8 × 96	0	8 × 8 × 96	0	8 × 8 × 96	0
BatchNorm2	8 × 8 × 96	384	8 × 8 × 96	384	8 × 8 × 96	384
Conv3	—	—	8 × 8 × 128	110720	8 × 8 × 128	110720
MaxPool3	—	—	4 × 4 × 128	0	4 × 4 × 128	0
BatchNorm3	—	—	4 × 4 × 128	512	4 × 4 × 128	512
Conv4	—	—	—	—	4 × 4 × 160	184480
MaxPool4	—	—	—	—	2 × 2 × 160	0
BatchNorm4	—	—	—	—	2 × 2 × 160	640
Flatten	6144	0	2048	0	640	0
FC1	1000	6145000	1000	2049000	1000	641000
FC2	100	100100	100	100100	100	100100
Output	10	1010	10	1010	10	1010
Total	—	6,303,934	—	2,319,166	—	1,096,286

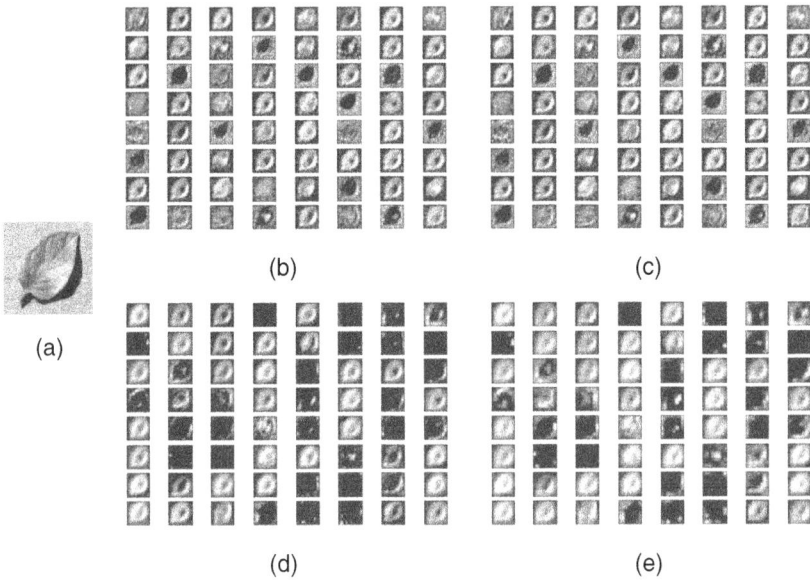

Figure 3.3 Feature maps. (a) Original diseased tomato leaf image (b–e). First 64 feature maps of all 4 conv. layers of custom-designed CNN with depth 4.

for multiclass classification. In this study, the down-sampling technique used is max pooling with a 2×2 filter size. Max Pooling forces the network to concentrate on a small number of neurons making a regularizing effect on the network, thereby reducing the overfitting problem. Here, the stochastic gradient descent algorithm is utilized for network training. Below is the learning rule for the stochastic gradient descent algorithm for a single epoch.

Step 1: *Load an input image X_i, forward pass it through the network and obtain the output Y_i*

Step 2: *Compute the soft margin loss $J(.)$*

Step 3: *if $J(.) < \varepsilon$ go to Step 7 else proceed with Step 4*

Step 4: *Compute $\nabla J(.) = \dfrac{\partial J(w)}{\partial y}$*

Step 5: *Compute ∇ net(.)*

Step 6: *Update the filter weights W (from output to input)*

Step 7: *Increment i, if i < No. of images,then go to Step 1 else go to Step 8*

Step 8: *End*

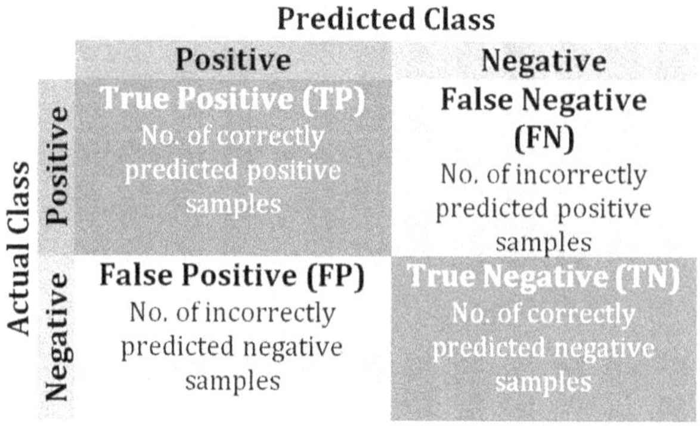

Predicted Class

		Positive	Negative
Actual Class	**Positive**	**True Positive (TP)** No. of correctly predicted positive samples	**False Negative (FN)** No. of incorrectly predicted positive samples
	Negative	**False Positive (FP)** No. of incorrectly predicted negative samples	**True Negative (TN)** No. of correctly predicted negative samples

Figure 3.4 Definitions of TP, TN, FP and FN.

3.2.4 Performance measures

3.2.4.1 Accuracy

The ratio of the number of accurate predictions to the total number of predictions made is widely employed as a performance metric for classification algorithms (Eqn. 3.4).

$$\text{Accuracy} = \frac{\text{TP} + \text{TN}}{\text{TP} + \text{FP} + \text{FN} + \text{TN}} \tag{3.4}$$

The details of TP, TN, FP and FN are shown in Figure 3.4.

3.2.4.2 Precision

Precision is the ratio between true positives (TP) and all predicted positives (TP+FP) and is shown in Eqn. 3.5.

$$\text{Precision} = \frac{\text{TP}}{\text{TP} + \text{FP}} \tag{3.5}$$

3.2.4.3 Recall

Recall is the ratio between true positives (TP) and all positives (TP + FN) and is given in Eqn. 3.6.

$$\text{Recall} = \frac{\text{TP}}{\text{TP} + \text{FN}} \tag{3.6}$$

3.2.4.4 F1 score

F1-score is actually the harmonic mean of the precision and recall and is given in Eqn. 3.7.

$$F1\,Score = \frac{2 \times Precision \times Recall}{Precision + Recall} \tag{3.7}$$

3.3 EXPERIMENTAL RESULTS AND DISCUSSION

First, the experiment is performed using a traditional machine learning model. The hand-crafted features used here are extracted using Log-Gabor filters. To extract the hand-crafted features, the leaf images are convolved with Log-Gabor filters of scale 2 and orientation 4 ([2S, 4O]), which results in the eight feature maps. From every feature map, features such as mean and standard deviation are calculated, resulting in 16 features [35]. The extracted features are fed into an SVM classifier for classification. SVM creates a model using training data and then predicts the target values of the test data; applications include object recognition [36–38], medicinal plant recognition [39] etc. In SVM, linearly separable data may be analyzed using a hyperplane, whereas non-linearly separable data are analyzed using kernel functions such as higher-order polynomials, Gaussian Radial Basis Function (RBF), and Tan-Sigmoid. The kernels used in the experimentation are linear and RBF kernels. The experimentation is repeated with Log-Gabor filters of scale 4 and orientation 8 ([4S, 8O]). The performance measures of the traditional machine learning model are depicted in Table 3.4.

From Table 3.4, it is inferred that Log-Gabor filter of scale 4 and orientation 8 provides better performance measures. The confusion matrix for the same is depicted in Figure 3.5. Next the experiment is carried out using a custom-made deep learning model. For experimentation purposes, the network parameters, learning rate $\eta = 0.01$ and mini-batch size =16 are considered. The optimizer used is RMSProp optimizer. The deep features are extracted using the three different custom-designed CNN architectures for different epochs. The accuracies achieved by the deep learning models are given in Table 3.5.

From Table 3.5, it is inferred that a custom-designed CNN architecture with depth 3 gives better accuracy at 75 epochs. The performance measures of these custom-designed CNN models for 75 epochs are provided in Table 3.6. Figure 3.6 depicts the confusion matrix and the receiver operating characteristics (ROC) for custom CNN architecture with depth 3. Figure 3.6 shows that the better ROC curve is obtained for the Curl Virus and Healthy categories, providing a maximum area of 0.99 under the curve.

Table 3.4 Performance measures of the traditional machine learning model

Performance measure	SVM kernel	Log-Gabor filter scale & orientation	Tomato leaf disease									
			TS	MV	CV	BS	EB	H	LB	LM	SLS	SM
Precision	Linear	[2S, 4O]	0.56	0.29	0.63	0.57	0.35	0.65	0.50	0.49	0.39	0.50
		[4S, 8O]	0.67	0.56	0.75	0.70	0.48	0.79	0.63	0.63	0.62	0.70
	RBF	[2S, 4O]	0.56	0.50	0.73	0.54	0.47	0.89	0.59	0.74	0.47	0.48
		[4S, 8O]	0.70	0.59	0.79	0.66	0.58	0.93	0.70	0.73	0.66	0.73
Recall	Linear	[2S, 4O]	0.41	0.06	0.79	0.75	0.16	0.77	0.54	0.16	0.30	0.58
		[4S, 8O]	0.66	0.53	0.85	0.83	0.27	0.85	0.63	0.39	0.58	0.70
	RBF	[2S, 4O]	0.40	0.14	0.82	0.85	0.26	0.87	0.55	0.29	0.30	0.73
		[4S, 8O]	0.72	0.46	0.85	0.87	0.43	0.95	0.67	0.42	0.56	0.79
F1-score	Linear	[2S, 4O]	0.47	0.10	0.70	0.65	0.22	0.71	0.52	0.24	0.34	0.53
		[4S, 8O]	0.67	0.55	0.79	0.76	0.35	0.82	0.63	0.48	0.60	0.70
	RBF	[2S, 4O]	0.47	0.22	0.77	0.66	0.33	0.88	0.57	0.42	0.37	0.58
		[4S, 8O]	0.71	0.51	0.82	0.75	0.49	0.94	0.68	0.54	0.61	0.76
Overall Accuracy (%)	Linear	[2S, 4O]	55.14									
		[4S, 8O]	68.24									
	RBF	[2S, 4O]	61.03									
		[4S, 8O]	**73.01**									

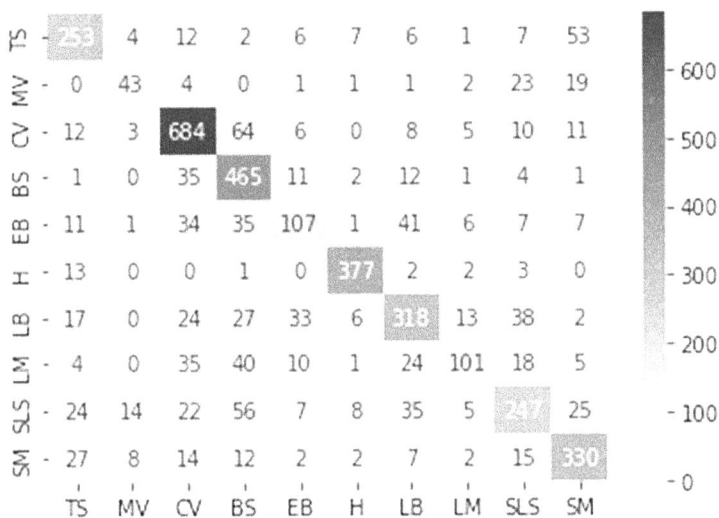

Figure 3.5 Confusion matrix for traditional machine learning model for Log-Gabor filters of scale 4 and orientation 8.

Table 3.5 Accuracy of the deep learning models

	Accuracy (%)		
Epochs	Depth 2	Depth 3	Depth 4
25	69.27	92.21	86.91
50	83.30	90.88	92.73
75	90.93	**94.00**	90.28

Table 3.6 Performance measures of the custom-designed CNN models for 75 epochs

Tomato leaf disease	Precision			Recall			F1-score		
	Depth 2	Depth 3	Depth 4	Depth 2	Depth 3	Depth 4	Depth 2	Depth 3	Depth 4
TS	0.8	0.94	0.68	0.9	0.85	0.93	0.86	0.89	0.79
MV	0.97	0.94	0.97	0.9	0.95	0.9	0.93	0.94	0.93
CV	0.98	0.98	1.00	1.00	0.99	0.96	0.97	0.98	0.98
BS	0.96	0.97	0.96	0.9	0.95	0.95	0.95	0.96	0.96
EB	0.78	0.74	0.91	0.6	0.9	0.61	0.69	0.81	0.73
H	0.98	1.00	0.91	1.00	0.98	1.00	0.98	0.99	0.95
LB	0.82	0.94	0.9	0.9	0.94	0.91	0.87	0.94	0.9
LM	0.92	0.89	0.93	0.8	0.95	0.86	0.86	0.92	0.89
SLS	0.87	0.98	0.84	0.9	0.87	0.97	0.9	0.92	0.9
SM	0.94	0.92	0.96	0.9	0.96	0.75	0.91	0.94	0.84

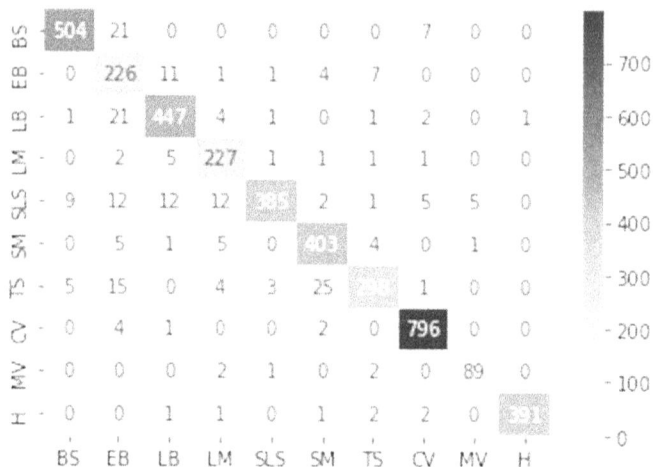

(a)

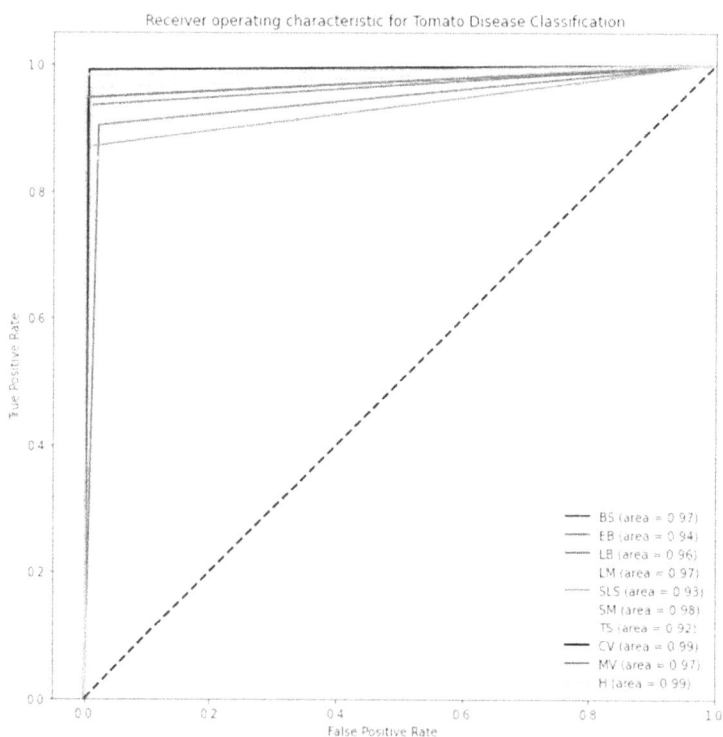

(b)

Figure 3.6 (a) Confusion matrix for custom CNN architecture with depth 3. (b) ROC curve.

3.4 CONCLUSION AND FUTURE WORK

In this work, we have compared the potential efficiency of the traditional machine learning model with the deep learning model for the problem of tomato leaf disease classification. We have achieved an overall accuracy of 73.01% for the traditional machine learning model which uses a Log-Gabor filter for the hand-crafted features. On the other hand, a custom-designed CNN architecture with depth 3, a deep learning model, provided an accuracy of 94% using the machine-learnt features. Experiments also show that the accuracy is increased by increasing the depth of the convolutional layers in the deep learning model. This work can be further extended to enable inexperienced farmers to make rapid informed judgments about tomato leaf disease by combining the trained model with mobile devices.

REFERENCES

1. Stilwell, M. The Global Tomato Online News Processing in 2018. Available online: https://www.tomatonews.com/
2. Schreinemachers, P., Simmons, E.B., and Wopereis, M.C. Tapping the economic and nutritional power of vegetables. *Global Food Security*, 2018, 16, 36–45.
3. Tian, Y. W. and Niu, Y. Applied research of support vector machine on recognition of cucumber disease. *Journal of Agricultural Mechanization Research*, March 2009, 31, 36–39.
4. Cui, Y. L., Cheng, P. F., Dong, X. Z., Liu, Z. H., and Wang, S. X. Image processing and extracting color features of greenhouse diseased leaf. *Transactions of the CSAE*. December 2005, 21(Supp.), 32–35.
5. Cen, Z. X., Li, B. J., Shi, Y. X., Huang, H. Y., Liu, J., Liao, N. F., and Feng, J. Discrimination of cucumber anthracnose and cucumber brown speck base on color image statistical characteristics. *Acta Horticulturae Sinica*, June 2007, 34, 1425–1430.
6. Zhao, Y. X., Wang, K. R., Bai, Z. Y., Li, S. K., Xie, R. Z., and Gao, S. J. Research of maize leaf disease identifying system based image recognition. *Scientia Agricultura Sinica*, April 2007, 40, 698–703.
7. Pydipati, R., Burks, T. F., and Lee, W. S. Identification of citrus disease using color texture features and discriminant analysis. *Computers and Electronics in Agriculture*, June 2006, 52, 49–59.
8. Ramya, V. and Lydia, M. A. Leaf disease detection and classification using neural networks. *International Journal of Advanced Research in Computer and Communication Engineering*, 2016, 5(11), 207–210.
9. Sujatha, R. et al. Performance of deep learning vs machine learning in plant leaf disease detection. *Microprocess. Microsystems*, 2021, 80, 103615.
10. Ahila Priyadharshini, R., Arivazhagan, S., Francina, E. C., and Supriya, S. Leaf Disease Detection And Classification System For Soybean Culture. In: *2019 1st International Conference on Innovations in Information and Communication Technology (ICIICT)*, Chennai, India, 2019, pp. 1–7, doi: 10.1109/ICIICT1. 2019.8741482.

11. Kantale, P. Pomegranate disease classification using Ada-Boost ensemble algorithm. *International Journal of Engineering Research & Technology*, 2020, 9(09), 612–620.
12. Johannes, A., Picon, A., and Alvarez-Gila, A., et al. Automatic plant disease diagnosis using mobile capture devices, applied on a wheat use case. *Computers and Electronics in Agriculture*, 2017, 138, 200–209.
13. Basavaiah, J. and Arlene Anthony, A. Tomato leaf disease classification using multiple feature extraction techniques. *Wireless Personal Communications*, 2020, 115(1), 633–651.
14. Kalyoncu, C. and Toygar, Ö. GTCLC: Leaf classification method using multiple descriptors. *IET Computer Vision*, 2016, 10(7), 700–708.
15. Kaur, S., Pandey, S., and Goel, S. Semi-automatic leaf disease detection and classification system for soybean culture. *IET Image Process*, 2018, 12(6), 1038–1048.
16. Banerjee, N., Sathish, R., and Sheet, D. Deep neural architecture for localization and tracking of surgical tools in cataract surgery. *Computer Aided Intervention and Diagnostics in Clinical and Medical Images*, 2019, 31, 31–38.
17. Tom, F. Debdoot Sheet, Simulating Patho-realistic Ultrasound Images using Deep Generative Networks with Adversarial Learning. In: *IEEE International Symposium on Biomedical Imaging*, Washington, DC, USA 2018, 1174–1177.
18. Kumar, H. and Ravindran, B. Polyphonic music composition with LSTM neural networks and reinforcement learning, arXiv preprint arXiv:1902.01973, 2019.
19. Madakannu, A. and Selvaraj, A. DIGI-Net: A deep convolutional neural network for multi-format digit recognition. *Neural Comput & Applic*, 2020, 32, 11373–11383. https://doi.org/10.1007/s00521-019-04632-9
20. Brahimi, M., Boukhalfa, K., and Moussaoui, A. Deep learning of tomato diseases: Classification and symptoms visualization. *Applied Artificial Intelligence*, 2016, 31(4), 1–17. https://doi.org/10.1080/08839514.2017.1315516
21. Trivedi, N. K., Gautam, V., Anand, A., Aljahdali, H. M., Villar, S. G., Anand, D., Goyal, N., and Kadry, S. Early detection and classification of tomato leaf disease using high-performance deep neural network. *Sensors*, 2021, 21, 7987. https://doi.org/10.3390/s21237987
22. Ouhami, M., Es-Saady, Y., Hajji, M. E., Hafiane, A., Canals, R., and Yassa, M. E. *Deep Transfer Learning Models for Tomato Disease Detection*. Image and Signal Processing. ICISP 2020. Lecture Notes in Computer Science(), vol 12119. Springer, Cham, 2020. https://doi.org/10.1007/978-3-030-51935-3_7
23. Yang, L., Yi, S., Zebg, N., Liu, Y., and Zhang, Y. Identification of rice diseases using deep convolutional neural networks. *Neurocomputing*, 2017, 267, 378–384.
24. Kawaski, R., Uga, H., Kagiwada, S., and Iyatomi, H. Basic study of viral plant diseases using convolutional neural networks. In: *Proceedings of the International Symposium on Visual Computing*, Las Vegas, NV, USA, pp. 638–645, 2015.
25. Ahila Priyadharshini, R., Arivazhagan, S., Arun, M., and Mirnalini, A. Maize leaf disease classification using deep convolutional neural networks. *Neural Comput & Applic*, 2019, 31, 8887–8895. https://doi.org/10.1007/s00521-019-04228-3.
26. Zaki, S. Z. M., Zulkifley, M. A., Mohd Stofa, M., Kamari, N. A. M., and Mohamed, N. A. Classification of tomato leaf diseases using mobilenet v2. IAES. *IAES International Journal of Artificial Intelligence*, 2020, 9(2), 290–296.

27. Agarwal, M., Singh, A., Arjaria, S., Sinha, A., and Gupta, S. ToLeD: Tomato leaf disease detection using convolution neural network. *Procedia Computer Science*, 2020.

28. Wang, L., Sun, J., Wu, X., Shen, J., Lu, B., and Tan, W. Identification of crop diseases using improved convolutional neural networks. *IET Computer Vision*, 2020, 14(7), 538–545.

29. Agarwal, M., Gupta, S. K., and Biswas, K. K. Development of efficient CNN model for tomato crop disease identification. *Sustainable Computing: Informatics and Systems*, 2020, 28, 100407.

30. Geetharamani, G., and Arun, P. J. Identification of plant leaf diseases using a nine-layer deep convolutional neural network. *Computers and Electrical Engineering*, 2019, 76, 323–338.

31. Nithish, E. K., Kaushik, M., Prakash, P., Ajay, R., and Veni, S. Tomato leaf disease detection using convolutional neural network with data augmentation. In: *Proceedings of the 5th International Conference on Communication and Electronics Systems ICCES 2020*, Coimbatore, India, pp. 1125–1132, 2020.

32. Mohanty, S. P., Hughes, D. P., and Salathé, M. Using deep learning for image-based plant disease detection. *Frontiers in Plant Science*, 2016, 7. https://doi.org/10.3389/fpls.2016.01419.

33. YuXiaoyu, Q., and TanHong, W. H. W. A Face Recognition Method Based on Total Variation Minimization and Log-Gabor Filter, *International Conference on Electromechanical Control Technology and Transportation*, Zhuhai City, Guangdong Province, China, 2015.

34. Field, D. Relations between the statistics of natural images and the response properties of cortical cells. *Journal of the Optical Society of America A*, 1987, 4, 12, 2379–2394.

35. Priyadharshini, R. A., Arivazhagan, S., and Sangeetha, L. Vehicle recognition based on Gabor and Log-Gabor transforms. In: *2014 IEEE International Conference on Advanced Communications, Control and Computing Technologies*, 1268–1272, Ramanathapuram, India, 2014. doi:10.1109/icaccct.2014.7019303

36. Priyadharshini, R. and Arivazhagan, S. Object recognition based on local steering kernel and SVM. *International Journal of Engineering*, 2013, 26(11), 1281–1288.

37. Ahilapriyadharshini, R., Arivazhagan, S., and Gowthami, M. Weber Local Descriptor based object recognition. In: *2012 IEEE International Conference on Advanced Communication Control and Computing Technologies (ICACCCT)*, Ramanathapuram, India, pp. 115–119, 2012. doi: 10.1109/ICACCCT.2012.63 20753.

38. Arivazhagan, S. and Ahila Priyadharshini, R. Generic visual categorization using composite Gabor and moment features, *Optik*, 2015, 126, 21, 2912–2916, https://doi.org/10.1016/j.ijleo.2015.07.035.

39. Ahila Priyadharshini, R., Arivazhagan, S., and Arun, M., Ayurvedic medicinal plants identification: A comparative study on feature extraction methods computer vision and image processing. CVIP 2020. *Communications in Computer and Information Science*, 2021, 1377. https://doi.org/10.1007/978-981-16-1092-9_23

Chapter 4

Application of artificial intelligence and automation techniques to health service improvements

Law Kumar Singh
GLA University, Mathura, India

Munish Khanna
Hindustan College of Science and Technology, Mathura, India

Rekha Singh
Uttar Pradesh Rajarshi Tandon Open University, Prayagraj, India

CONTENTS

DOI: 10.1201/9781003415466-4

4.1 INTRODUCTION

Disease, multimorbidity, and disability are now more prevalent as a result of aging and demographic change. Rising social expectations, a rise in health-care demand, and an increase in the cost of healthcare services all have an effect on this tendency. The inefficiency of the system, resulting in low levels of output, is another challenge to overcome [1]. The current climate of fiscal conservatism, together with misguided economic austerity measures, is limiting investment in the infrastructure of the healthcare system [2].

According to the World Health Organization (WHO), in order to overcome these challenges and achieve UHC by the year 2030, the current method of dispensing medical care will need to undergo a significant transformation. Machine learning, now in its infancy, is the most visible manifestation of artificial intelligence, as well as the most current research area in digital technology, and importantly, it has the potential to allow us to accomplish more with fewer resources in the future [3].

While AI is crucial to this transformation [4], there is currently no hard evidence to prove the extent to which digital technology is altering health-care delivery systems. Will AI operate well in this setting or will it fall short, much like earlier attempts to incorporate digital technology into the work-place and society at large? This study focuses on the potential applications of AI in healthcare systems, and how these systems could change as a consequence. One way to achieve the goal of UHC is to make public health and healthcare systems more equitable, responsive, universal and cost-effective.

4.2 THE ADVANCE OF ARTIFICIAL INTELLIGENCE AND MACHINE LEARNING

Artificial intelligence (AI) is a wide topic that spans many different subfields. An AI system is one that has the ability to learn from data and apply that learning to new situations, a feature commonly associated with human intellect. AI attempts to investigate and develop systems that exhibit intelligence-like traits such as reasoning and problem solving without the requirement for human intelligence. AI systems are becoming more and more sophisticated with each passing day. In spite of its shortcomings, AI is widely used, and overlaps in some ways with modern statistical methodologies. Whether or not this is a good thing is up for discussion.

The theoretical underpinnings of AI research can be found in many academic disciplines, including computer science, philosophy, and mathematics. Understanding and developing systems with intelligence-like qualities is the ultimate goal of artificial intelligence. The recent acceleration of progress in the field of AI can be attributed to a number of approaches, including deep learning [5]. Deep learning, which was first proposed and pioneered by researchers at Google, is an AI discipline that integrates a variety of learning approaches, including reinforcement learning. As a subfield of machine learning, it encompasses a wide range of methodologies, including convolutional neural networks (CNNs), convolutional neural networks with embedded recurrent neural networks, and others. Deep learning enables computers to recognize associations in large volumes of raw data and then apply those relationships to real-world scenarios. One example is computers' ability to recognize patterns in digital photographs based on the individual pixels that make up the images. Another example is the use of deep learning to create a digital assistant capable of recognizing faces. Deep learning is currently being used in medicine to aid in the discovery of previously unknown linkages and correlations. Many companies across many industries have been using deep learning algorithms for a long time in a range of applications [6]. Several sectors have already created new guidelines for the use of deep learning systems, and this can be expected to continue. In short, deep learning algorithms have raised the bar and are setting new norms in business sectors with a wealth of high-quality digital data and a strong economic incentive to automate prediction activities. These include healthcare, retail, and financial services.

4.2.1 Machine learning

Machine learning teaches computer programs, or algorithms, to predict the future accurately by looking at prior data and comparing it to those predictions. It is built on the principles of statistical modeling and computer-assisted data analysis. Machine learning employs statistical techniques that are significantly more extensive than those that are currently typically utilized in medicine. Reinforcement learning and other newer approaches like deep learning are used to analyze more complex data since they make fewer assumptions about the underlying data and hence can process more data.

4.2.2 Deep learning

A vast amount of raw data can be fed into a deep learning machine to enable it to learn more about its surroundings, and the system can subsequently generate the representations required for detection or classification. Deep learning algorithms are designed to improve the most relevant parts of the input while suppressing less significant variations in the classification process by using different representations of the data. This is performed by utilizing the data in a way that allows for many interpretations at the same time. Supervision is not required for deep learning to take place. A number of the most notable improvements in machine learning have been made using deep learning techniques.

4.2.3 Supervised learning

Computer algorithms are trained by studying the outputs of interest selected by a (usually human) supervisor. Future occurrences can be predicted through associative thinking. There are various examples of machine learning in the healthcare and other industries.

4.2.4 Unsupervised learning

Automated algorithms can find patterns in datasets without the need for an external description of the important associations. Such algorithms can find previously unknown predictions rather than relying primarily on established associations.

4.2.5 Reinforcement learning

It is possible for computers to learn by maximizing a predetermined reward. Making a mistake in a video game has no real-world repercussions because the data is perfect and the possibilities are limitless – a strategy largely inspired by behavioral psychology.

Humans' ability to curate large amounts of data and optimize deep learning algorithms has led to the discovery of correlations with high predictive value, frequently for a specific use case. In order for machine learning to combine concepts learned independently, much as the human brain does, it is necessary to first establish a self-reinforcing loop. Machine learning can then be applied in a variety of contexts and settings.

4.3 THE USE OF AI IN THE HEALTHCARE DECISION-MAKING PROCESS

Many data-processing tasks must be completed if public health systems are to function efficiently. In order to attain the desired system outcomes, policymakers must make adjustments to the structure and governance of health systems, as well as to financial and resource management [8].

Screening and diagnosis (the classification of cases based on history, examination and investigations) and treatment planning, execution, and monitoring are the two core information-processing responsibilities in the delivery of health care. Hypothesis development, hypothesis testing, and subsequent action are all based on these fundamental concepts. The ability of machine learning to discover previously unseen patterns in data can greatly assist healthcare systems.

Machine learning techniques can detect patterns in data without making any assumptions about how the data is distributed [9]. A machine learning model with more variables and data types, on the other hand, is more difficult to grasp, but it is also more generalizable and capable of responding to more intricate situations. In various studies, these strategies have been used to screen for, diagnose, and predict the occurrence of probable problems in the future. Because of the limitations of a single data center, these installations cannot be replicated or generalized [11]. For example, a higher proportion of patients in healthcare facilities are at risk of contracting influenza than in the general population [12].

4.4 THE POTENTIAL IMPACT OF AI ON THE HEALTHCARE WORKFORCE AND CLINICAL CARE

The fact that data handling is ubiquitous in the healthcare industry and other sectors of society shows that machine learning is progressing at a rapid pace [13]. With widespread use, room for improvement over time, and the potential to spark additional research, machine learning has become what some have dubbed a "general purpose technology" [14]. The introduction of such technologies typically causes "widespread economic disruption" and creates both winners and losers. The economists Acemoglu and Restrepo assert that as industries become more competitive, people will be replaced by machines. As businesses become more profitable and efficient, a productivity benefit reduces this displacement effect [15]. The money saved can then be used to develop new non-automatable jobs and keep non-automatable processes up and running; however, some of these will necessitate working directly on the automating technology.

Consider diagnostic radiology, a clinical topic that is already widely covered in machine learning literature, to exemplify how this general tendency might apply to the health care profession [16]. Some doubters have predicted the demise of the radiology profession and questioned the value of continuing to train new generations of radiologists in light of the fact that deep learning algorithms have redefined the possibilities of diagnostic image processing. As non-radiologists gain more autonomy, machine learning techniques may be deployed increasingly frequently to help them with diagnostic image processing. As a direct result of this breakthrough, radiologists in practice will have more time to spend with new patients. Healthcare organizations would be able to rebalance the skill mix and dispersion of radiology teams thanks to this change in employment. This would be achievable because more work would be done at the primary care level, and fewer radiologists would be needed to handle work that could not be automated and less common situations in secondary and tertiary hospitals. A machine learning system first "reads" the image and indicates regions that the human radiologist should pay special attention to (an early example was the system responsible for diagnosing pneumonia). This improves the efficiency of the workflow by enabling a human decision maker to direct their limited attention to the areas in which it can be utilized most effectively, while simultaneously managing a greater number of scenarios. Similar uses are likely to have an effect on pathology as well as other image-based scientific disciplines. The result will be the development of machine learning approaches that mix human and machine intelligence [17]. Human ability to produce ideas can be combined with AI's ability to analyze massive volumes of data to identify expected correlations or improve against a success criterion. It has been suggested that radiology and pathology could be integrated into a new profession called "information expert", whose job will be to manage information obtained by artificial intelligence in the context of the patient's therapeutic setting.

4.5 CREATING A SUPPORTIVE ENVIRONMENT FOR THE USE OF AI IN HEALTH SYSTEMS

Machine learning-related technology is evolving swiftly. Even with significant scientific developments, machine learning in the field of health will be hampered in the future owing to the lack of an environment encouraging its use and development, as is the case with other emerging technologies. A receptive context for AI requires, among other things, the availability of curated data, a supportive legal environment, legal protections for citizens' rights, clear guidelines for accountability, and the ability to manage strategic change.

4.6 ORGANIZATION OF DATA

Many healthcare companies have experienced interoperability issues as a result of inconsistent data structures. More data is needed for machine learning than can be handled by traditional healthcare methods (s24). Google worked with three academic medical institutions to conduct an unannounced investigation to determine whether it was possible to combine data from three teaching hospitals. In this study, deep learning models beat previous benchmarks in properly projecting all patient final diagnoses, 30-day unplanned readmissions, lengthier hospital stays, and in-hospital mortality (s25). Advances in machine learning such as these do not eliminate the requirement for data harmonization throughout a health system. Data aggregation tasks must be managed by health systems in order for them to work normally and support machine learning. This project needs much greater attention in view of the potential advantages of machine learning.

4.7 TRUST AND DATA MANAGEMENT

The lack of trust in the use of data constitutes a substantial obstacle to the collection of data sets required for the development of machine learning systems. Recent incidents of overzealous data sharing that broke legal constraints have undermined trust between those who utilise data and the citizens whose information is being exploited. Although it is widely accepted in the consumer digital economy that individuals submit their data in exchange for more relevant search results or social network feeds, it is uncertain if patients share this implicit understanding. This may be the case if patients, unlike retail customers, simply do not understand the benefits of data sharing. The retail and healthcare industries also differ qualitatively, and any possible long-term benefits of data interchange are outweighed by consumers' fundamental concerns about their privacy, confidence in governments, or fear of getting varied treatment depending on their health status.

4.8 WORKING WITH THE TECHNOLOGY INDUSTRY

Worryingly, advances in machine learning will come from or require collaboration with a small set of IT companies that have already spent billions of dollars developing the processing, storage, and intellectual resources that are required for machine learning. The fact that the policies of national governments are becoming more complicated as economic concentration develops highlights more than ever the relevance of increased economic concentration. As a result, it's likely that a select few private technology companies will eventually be responsible for providing the necessary AI infrastructure. In order to work with these for-profit technology companies, allow for the widespread collecting and use of health data, preserve privacy, and provide equal compensation for created intellectual property, new contractual approaches are necessary. So far these are lacking, as is the appropriate guidance. Private businesses and international public health organizations have a fantastic chance to demonstrate leadership in ensuring that the advantages of AI are widely disseminated in the absence of an agreed-upon framework for contracts and intellectual property rights. These decisions present an opportunity to promote social cohesion and demonstrate corporate social responsibility.

4.9 ACCOUNTABILITY

Deep neural networks, a type of neural network that is often called a "black box", need millions of data points to create models that can classify thousands of things. This is because they often use data in ways that make internal representations that are hard for humans to understand. Because of this, it becomes very hard to explain how a conclusion drawn from data works in the context of standard statistical models. Under the European Union's General Data Protection Regulation, a person will be able to find out more about a decision that was made about them using "automated processing". Decisions with unclear reasons will be closely looked at.

A research paper titled "Responsibility of AI Under the Law" has recently been published by the Berkman Klein Working Group on Explanation and the Law at Harvard University. In this paper, the authors present the findings of an investigation into the question of how machine learning systems ought to be held responsible. They take into account the particulars of the issue that the machine learning system is intended to solve, and then they devise solutions to ensure that the system is able to live up to the standards that have been set for it. It is possible that theoretical limitations or statistical data from trials of machine learning systems are all that is required to solve certain problems with more explicit definitions. On the other hand, being responsible in clinical practice necessitates providing an explanation, which frequently entails amorphous objectives and the presence of a number of factors that are irrelevant. For an observer to figure out how much a single input determined or was significant to an outcome, an explanation is required. The authors argue that it is fair to expect an AI system to explain itself in scenarios in which a person would explain things to themselves. The accomplishment of this goal, however, requires investing a significant amount of money in technology in order to create deductive reasoning systems, specific to the organization, that are able to describe the operation of both machine learning algorithms and deep learning algorithms.

4.10 MANAGING STRATEGIC TRANSFORMATION CAPABILITY

The unexpected impacts that machine learning will have on the workforce, both inside and outside the healthcare industry, are of concern to policymakers. Machine learning-based diagnosis, care management, and monitoring will not be used in practice until there is proof that they are more effective than the current algorithms. For medical experts and government authorities to have faith in machine learning-based systems, they must provide results that are practical in the real world. This can be done by performing experimental testing or by assessing how well the systems work in the real world. Previously finished projects may need to be redone as a result of the improvements made to algorithms due to new machine learning techniques and improved accessibility to data. This may face health care systems with huge expenses that would need to be covered by increases in labor productivity if they are to continue operating sustainably.

In this scenario, the "displacement effect" will be felt most keenly by those working in less skilled manual and non-manual jobs. More people may be looking for work in healthcare systems that already have a sufficient number of healthcare professionals, particularly in roles involving psychological and emotional well-being, as well as caring for the elderly and disabled, which are frequently regarded as skilled and incapable of being automated. It is expected that the use of machine learning algorithms will enable elderly populations to be provided with superior community-based care and chronic illness management, while at the same time the number of front-line healthcare staff can be increased. Governments will need to take the lead and invest in reskilling "displaced" workers, training them for other options, such as new careers in the production and curation of data sets and machine learning algorithms. However, this may not be enough to offset the broader consequences of automation on labor markets.

4.11 AI-BASED BIOMETRIC AUTHENTICATION

Traditional methods of authentication include using passwords that are made up of a combination of letters, numbers, and symbols. Systems that need "something you know and something you have" are also regarded as classic methods of authentication [26, 27]. The use of these systems runs the risk that you might forget something, lose something, or have something taken from you. Traditional means of identification are rapidly becoming obsolete with the increasingly ubiquitous authentication based on a person's biometric traits. One of the more recent forms of biometrics that have been investigated is the electrocardiogram, sometimes known as an ECG. Current research presents an ECG-based authentication system that has the potential to be utilized not just for safety checks but also in therapeutic contexts. Researchers who are studying ECG-based biometric identification systems may find the proposed method helpful in determining the limits of their datasets and producing high-quality training data. We investigated the usefulness of this system and found that it probably achieves a success rate of around 92% in recognizing individuals.

4.12 RESULTS

4.12.1 Application of AI in diabetic detection using machine learning

Insulin resistance and a drop in insulin production by the pancreas are at the root of type 2 diabetes (T2D), which shows up as high blood glucose levels [18]. The goal of this study is to find traits that are linked to T2D that can be used to diagnose and treat the disease. The Pima Indian Diabetes Dataset, which was found by entering the Kaggle ML competition, was used to divide people with type 2 diabetes into groups. After the data was cleaned up, subsets of features were made by picking and grouping individual features in the

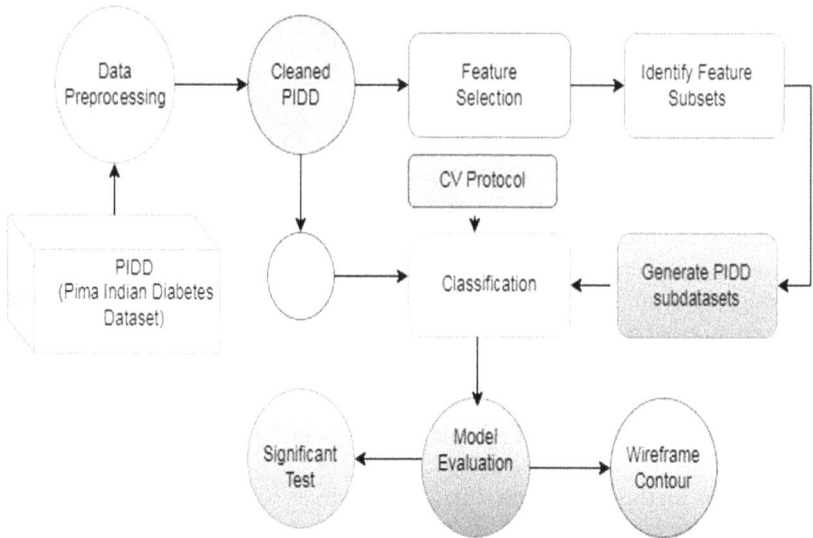

Figure 4.1 Diabetic retinopathy and AI-based model [18].

right way. Figure 4.1 shows the accuracy, kappa statistics, AUROC, sensitivity, specificity, and logarithmic loss (logloss) investigated for each candidate. We used summary statistics and a resampling distribution to determine how well the classifiers worked. The model with the highest level of accuracy is the generalized boosted regression model (90.91%), followed by the kappa model (78.77%), and then by the specificity model (85.19%). Researchers have found that looking at factors like age, diabetes pedigree function, body mass index (BMI), and glucose levels are the best ways to predict how things will turn out. Based on these results, it seems that machine learning approaches may be able to improve prediction models for type 2 diabetes and help the Pima Indian population establish which factors affect outcomes.

4.12.2 Detection of glaucoma using machine learning

This paper proposes a model for diagnosing glaucoma, using deep image analysis to look for signs of glaucoma in the retinal fundus [19, 23–25]. The parameters extracted include the ratio of the cup to the disk and the area of the inferior, superior, nasal, and temporal regions (Figure 4.2). In the model, four machine learning algorithms work together to give a classification accuracy of 98.6%. Other models, like support vector machine (SVM), K-nearest neighbors (KNN), and Naive Bayes, have accuracy rates of 97.61%, 90.47%, and 95.233%, respectively. These results show very clearly that the proposed model is the best way to look for early signs of glaucoma in the retinal fundus. The machine learning technology can also be used to detect glaucoma in OCT images [22].

Figure 4.2 Glaucoma detection in fundus image using machine learning.

4.12.3 Deep learning for glaucoma detection

The optic nerve, which conveys images of the outside world to the brain, is damaged by glaucoma [20]. This chronic infection is the second-largest cause of total blindness worldwide and if not treated promptly, results in a severe reduction in quality of life. Conventional methods for diagnosing glaucoma demand extensive expertise and expensive tools, making it challenging to evaluate numerous individuals at the same time. Costs are high and wait times are lengthy. Previously, artificial intelligence could not detect glaucoma because features had to be manually removed, a time-consuming and tedious task, but the development of deep learning (DL) techniques has enabled automatic feature extraction. Among the numerous issues that come with glaucoma are limited labeled data, the difficulty and expense of building glaucoma fundus photographic datasets, and the need for specialized hardware. Instead of a rigid DL model that only performs well on a single dataset and ignores these other issues, this study aimed to demonstrate a flexible DL model that performs well on a variety of datasets and can be applied in the real world. In this empirical investigation, two groups of fundus images – normal and glaucomatous – are distinguished using various deep learning algorithms. The publicly accessible ACRIMA, ORIGA, and HRF benchmark datasets were utilized for both training and testing. Not only are these models enhanced by transfer learning, but numerous others are as well. The models were validated using a total of 12 unique combinations of the aforementioned datasets, DRISHTI-GS, and a private dataset. Extensive testing was conducted to assess whether the proposed technique is effective. In terms of computed accuracy and area under the curve, the Inception-ResNet-v2 and Xception models surpass competing models. In the long term, ophthalmologists may gain a great deal from adopting computerized technologies to diagnose their patients (see Figure 4.3).

4.12.4 Deep learning techniques for glaucoma detection using OCT images

Glaucoma prediction was rendered simpler and more precise in an experiment combining deep learning with optical coherence tomography (OCT). The experiment predicts how glaucoma will develop using eight distinct

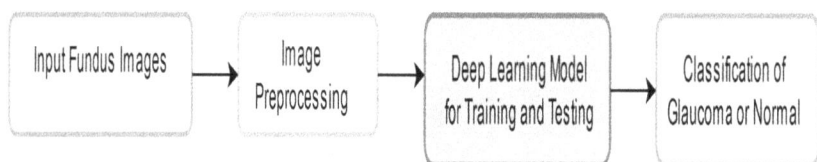

Figure 4.3 Glaucoma detection in fundus image using deep learning.

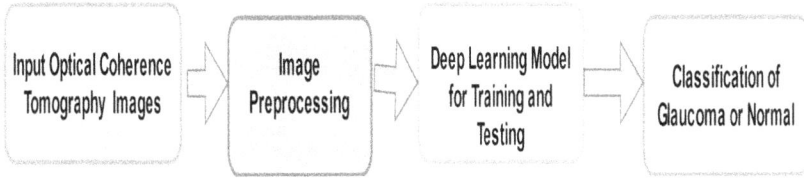

Figure 4.4 Glaucoma detection in OCT image using deep learning technique.

ImageNet models [21]. A variety of efficiency factors must be taken into account when comparing the effectiveness of several models. For each individual network, the performance of three distinct optimizers – Adam, Root Mean Squared Propagation, and stochastic gradient descent – was evaluated. Using transfer learning and fine-tuning techniques, the models' overall performance has increased. The initial training and evaluation involved a total of 4220 private photos (2110 normal OCT and 2110 glaucoma OCT). The models were evaluated using the well-known public standard dataset Mendeley. The VGG16 method was put to the test using the Root Mean Squared Propagation Optimizer, and the results demonstrate that it performs as intended with an overall accuracy of 95.68%. Various ImageNet models were found to be an effective alternative, since combining deep learning with the OCT modality makes glaucoma prediction simpler and more accurate. The study concluded that several ImageNet models should be used as a backup by computer-based automatic glaucoma screening systems (Figure 4.4).

One of the findings of the study was that glaucomatous OCT can be distinguished from conventional OCT using this totally automated method. Applying this procedure to people who are suspected of having this retinal infection not only makes it simpler to receive a proper diagnosis, which in turn helps prevent blindness, but also frees up time for senior ophthalmologists. The ability to distinguish between glaucoma OCT and conventional OCT automatically using this totally automated technology has great potential.

4.12.5 Conclusions

Ophthalmology is the principal focus of attention for researchers working with the forms of artificial intelligence that are now available. Using methods of machine learning, a variety of retinal illnesses can be identified from the corresponding photographs. This chapter examined artificial methods being implemented inside the healthcare system and described the diagnosis of glaucoma using machine learning techniques. While the chapter has looked at the direct influence that machine learning has had on health care systems, we have not considered the indirect effects that machine learning has had on the fundamental sciences.

REFERENCES

[1] Atun R. Transitioning health systems for multimorbidity. *Lancet* 2015;386: 721–2. doi:10.1016/S0140-6736(14)62254-6.

[2] Kocher R, Sahni NR. Rethinking health care labor. *New England Journal of Medicine* 2011;365:1370–2. doi: 10.1056/NEJMp1109649.

[3] Badawi O, Brennan T, Celi LA, Feng M, Ghassemi M, Ippolito A, et al. Making big data useful for health care: A summary of the inaugural mit critical data conference. *JMIR Medical Informatics* 2014;2:e22. doi: 10.2196/medinform.3447.

[4] Jones SS, Heaton PS, Rudin RS, Schneider EC. Unraveling the IT productivity paradox—lessons for health care. *New England Journal of Medicine.* 2012;366:2243–5. doi: 10.1056/NEJMp1204980.

[5] Beam A, Kohane I. Big data and machine learning in health care. *JAMA* 2018;319: 1317–8. doi: 10.1001/jama.2017.18391.

[6] LeCun Y, Bengio Y, Hinton G. Deep learning. *Nature* 2015;521:436. doi: 10.1038/nature14539.

[7] Marcus G. Deep learning: A critical appraisal. arXiv:1801.00631. 2018.

[8] Atun R, Aydın S, Chakraborty S, Sümer S, Aran M, Gürol I, et al. Universal health coverage in Turkey: Enhancement of equity. *Lancet* 2013;382:65–99. doi: 10.1016/S0140-6736(13)61051-X.

[9] Henglin M, Stein G, Hushcha PV, Snoek J, Wiltschko AB, Cheng S. Machine learning approaches in cardiovascular imaging. *Circ Cardiovasc Imaging* 2017;10:e005614.

[10] Johnson AE, Pollard TJ, Mark RG. Reproducibility in critical care: A mortality prediction case study, in *Proceedings of the 2nd Machine Learning for Healthcare Conference.* JMLR W&C Track Volume 68, 2017, November.

[11] Celi LA, Moseley E, Moses C, Ryan P, Somai M, Stone D, et al. From pharmacovigilance to clinical care optimization. *Big Data* 2014;2:134–41. doi: 10.1089/big.2014.0008.

[12] Brynjolfsson E, Mcafee AN. The business of artificial intelligence. Harv Bus Rev. 2017.

[13] Helpman E, Trajtenberg M. Diffusion of general purpose technologies. National Bureau of Economic Research. 1996. No. w5773.

[14] Trajtenberg M. AI as the next GPT: A political-economy perspective. National Bureau of Economic Research. 2018. No. w24245

[15] Acemoglu D, Restrepo P. Artificial intelligence, automation and work. National Bureau of Economic Research 2018. No. w24196.

[16] Siddartha M. The algorithm will see you now. *New Yorker* 2017;93:46–53.

[17] Howlader KC, Satu M, Awal M, Islam M, Islam SMS, Quinn JM, Moni, MA. Machine learning models for classification and identification of significant attributes to detect type 2 diabetes. *Health Information Science and Systems* 2022;10(1):1–13.

[18] Singh LK, Pooja, GH. et al. An enhanced deep image model for glaucoma diagnosis using feature-based detection in retinal fundus. *Medical & Biological Engineering & Computing* 2021;59:333–53. https://doi.org/10.1007/s11517-020-02307-5

[19] Singh LK, Pooja, GH. et al. Deep learning system applicability for rapid glaucoma prediction from fundus images across various data sets. *Evolving Systems* 2022. https://doi.org/10.1007/s12530-022-09426-4.

[20] Singh LK, Pooja, GH. et al. Performance evaluation of various deep learning based models for effective glaucoma evaluation using optical coherence tomography images. *Multimedia Tools and Applications* 2022;81:27737–81. https://doi.org/10.1007/s11042-022-12826-y

[21] Singh LK, Garg H, Khanna, M. An artificial intelligence-based smart system for early glaucoma recognition using OCT images. *International Journal of E-Health and Medical Communications (IJEHMC)* 2021;12(4):32–59.

[22] Singh LK, Garg, H. Detection of glaucoma in retinal images based on multiobjective approach. *International Journal of Applied Evolutionary Computation (IJAEC)* 2020;11(2):15–27.

[23] Singh LK, Garg, H. Detection of glaucoma in retinal fundus images using fast fuzzy C means clustering approach. In *2019 International Conference on Computing, Communication, and Intelligent Systems (ICCCIS)*, pp. 397–403. IEEE, 2019, October.

[24] Singh LK, Khanna M. A novel multimodality based dual fusion integrated approach for efficient and early prediction of glaucoma. *Biomedical Signal Processing and Control* 2022;73:103468.

[25] Singh LK, Khanna M, Thawkar, S, Gopal, J. Robustness for authentication of the human using face, ear, and gait multimodal biometric system. *International Journal of Information System Modeling and Design (IJISMD)*, 2021;12(1):39–72.

[26] Singh, LK, Khanna, M, Garg, H. Multimodal biometric based on fusion of ridge features with minutiae features and face features. *International Journal of Information System Modeling and Design (IJISMD)* 2020;11(1):37–57.

[27] Al Alkeem, E. et al. An enhanced electrocardiogram biometric authentication system using machine learning. *IEEE Access* 2019;7:123069–75, doi: 10.1109/ACCESS.2019.2937357.

Chapter 5

Artificial intelligence in disaster prediction and risk reduction

Siby Samuel
St. Aloysius (Autonomous) College, Jabalpur, India

CONTENTS

5.1 INTRODUCTION

A disaster is an occurrence that causes damage to the environment and the society in which we live, and hence has a significant impact on human life. Any type of crisis necessitates a fast response as well as a plan of action. Management is needed to protect lives, preserve infrastructure, and conserve resources.

Over the course of the last few years, there has been a discernible rise in the frequency of catastrophic natural disasters, leading to a substantial rise in the number of lives lost, as well as extensive damage to both the natural world and to human societies.

Compared to the preceding two decades (2000–2019), the year 2020 had a greater impact, both in terms of the number of events that were recorded and the economic losses that were incurred. In the year 2020, EMDAT recorded 389 natural disasters, resulting in the loss of 15,080 lives, affecting 98.4 million people, and costing 171.3 billion US dollars [1].

The magnitude of these losses implies an urgent need for enhanced risk management of natural hazards. In general, the management of natural disasters has always included a significant communication component. This component manifests itself in the form of alerts, warnings, notifications to citizens about changes in risks, evacuation orders, and other similar directives.

5.2 DISCOVERING THE UN-PREDICTING

Acute natural disasters, whether biological, geophysical, coastal, or related to hydrography or airspace, are capable of wreaking havoc on human society as well as the natural world and even further afield. These kinds of occurrences have a disproportionately negative influence on specific places (such as the least developed countries) and their inhabitants. The advent of artificial intelligence has made it possible to anticipate and mitigate natural disasters.

Snow avalanches are a type of natural catastrophe that may have serious repercussions on socioeconomic and geomorphic processes (Figure 5.1). These repercussions may take the form of damage to ecosystems, flora, landscape, infrastructures, and transportation networks, as well as human lives [2]. Using datasets that contain information about the weather, AI systems can be trained to make accurate predictions about the occurrence of avalanches [3]. There are a wide variety of applications for classification algorithms. Because the avalanche dataset contains a large number of data points

Figure 5.1 Snow avalanches damage ecosystems, infrastructures, transportation networks, and human lives. (Courtesy by https://www.itu.int/hub/wp-content/uploads/sites/4/2021/03/AI-to-understand-natural-disasters.jpg).

as well as a large number of factors, it is necessary to preprocess the dataset before attempting to determine the co-relationship that exists between the characteristics. This will allow for a better understanding of how the variables affect prediction.

5.2.1 Predicting shaking

A technique based on artificial intelligence (AI) may provide predictions about how earthquakes would shock places. AI can be used to create an earthquake early warning system that can anticipate how the ground will move during an earthquake and alert people several seconds before it begins [4] (Figure 5.2). To better anticipate the path of a future earthquake, deep neural network AI can look for patterns in previous quakes. In earthquake-prone areas, this might speed up processing and make it simpler to generalize.

5.2.2 AI in flood prediction

Around the world, devastating floods are making headline news. Floods are by far the most common natural catastrophe to occur, wreaking havoc on the environment, people's lives, and the economy (Figure 5.3), and causing the evacuation of hundreds of millions of people and the deaths of tens of thousands (based on typhoon hydrological and inundation data 2004–2015) [5]. Depending on the course of the river, satellite photos may be used to generate models of areas that are most at risk of flooding, and people can receive flood alerts in advance. AI approaches mimic the non-linear behavior of a given phenomenon, and are remarkably successful making predictions about the future. AI can foresee floods because of the vast quantities of high-quality datasets it has access to.

Figure 5.2 Damage to property and lives due to intense shaking of the ground (Courtesy: https://www.theguardian.com/world/2015/apr/25/science-of-earthquakes).

Figure 5.3 Flood causing economic and social damage (Courtesy: https://theconversation.com/people-couldve-prepared-for-the-floods-better-if-the-impacts-of-weather-forecasts-were-clearly-communicated-178309).

When disasters occur, an enormous quantity of information is compiled by a variety of sources, including but not limited to government agencies, emergency responders, loss claims adjusters, and social media [6]. However, there has been little research on how these data can be used to understand how different stakeholders are or can be directly or indirectly affected by large-scale natural disasters before, during, and after the event itself. Artificial intelligence (AI) systems that use large-scale disaster data in the real world and offer helpful tools for disaster forecasting, impact assessment, and societal resilience are becoming increasingly popular and necessary [7]. This will help with resource allocation, which can improve preparedness and prevention of natural disasters, save lives, reduce the negative impact on the economy, improve emergency response, and make communities stronger and more resilient. Using AI as a tool can reduce the risk of death, harm to the environment, and the impact on society, and help us respond to crises more effectively.

The practice of efficiently anticipating and responding to calamities is known as disaster management [8]. Mitigation, preparation, response, and recovery are the four separate phases that have been extensively identified [7, 8]. The term 'mitigation' refers to any action taken to prevent the occurrence of a disaster or to lessen the severity of the damage it causes. The term 'preparedness' refers to a variety of actions that communities can take to better equip themselves to respond to disasters or to lessen the effects of those disasters. These actions include emergency planning, supply stockpiling, training, and community education. The implementation of the plans that have been developed to protect life and property, as well as the environment and the socioeconomic structure of the neighborhood [9], is part of the response. The provision of emergency rescue and medical care, the opening and management of shelters, the distribution of supplies, and the evaluation of damage are all examples of the types of activities that fall under the umbrella of disaster relief and response. It is one of the most investigated stages, because this is the phase in which the people and the infrastructure require the most immediate assistance. In this phase of the process, efficiency is of the utmost importance; consequently, the techniques that are utilized place an emphasis not only on highly accurate results but also on quick and efficient methods. Recovery, also known as reconstruction, consists in the long-term measures that are taken in order to bring back a sense of normalcy to the community. During this stage, some of the activities that take place include providing financial support and construction or reconstruction (e.g., of buildings and key infrastructure). In addition, dynamic participation in disaster management can help local communities become more resilient [10].

5.3 DISASTER INFORMATION PROCESSING USING AI

5.3.1 Artificial intelligence in early warning system monitoring and disaster prediction

Artificial intelligence is capable of examining data to predict potential natural disasters ahead of time, sparing lives in the process. Early warning systems can be developed with the assistance of artificial intelligence (AI) in order to effectively and efficiently disseminate pertinent information about disaster events, in the form of alarms or warnings, to vulnerable communities either before or during a disaster. This paves the way for the adoption of proactive and preventive measures that can reduce the loss and damage caused by such events [11]. The goal of the system is to build prediction models that will allow for early decision-making, and that will, first and foremost, help reduce the number of lives lost as well as the damage done to the environment and the economy. Every conceivable kind of data, including that gleaned from analogous catastrophes and everyday situations, is compiled for the purpose. As a result of the growth of the Internet of Things and technology-driven sensor devices, it is now possible to create large amounts of data in a relatively short amount of time. EWS saves and analyses these data because they contain useful indicators and have the potential to open up many opportunities for monitoring and controlling both natural and man-made disasters. The early warning system (EWS) can mine early warning signals using artificial intelligence (AI), allowing proactive and preventive disaster mitigation, preparedness, response, and recovery measures to be planned, resulting in timely alerts and warnings being distributed to the relevant stakeholders.

5.3.1.1 Example

1. A flood prediction project can foresee torrential rains that will cause the rivers to rise and damage towns across the nation. Based on historical relevant data, AI aims to forecast potential rainfall as well as the potential rise in river levels, alerting the appropriate authorities to take the necessary measures.
2. Deep learning technology makes it feasible to reliably predict the occurrence of earthquakes and gain sufficient lead time for early warning systems. It is possible to predict the location of future earthquakes by analyzing previous seismic activity, which is depicted on topographic maps and in photographs. Using this information for training and testing, a deep learning network model was developed that is capable of reliably forecasting earthquakes in the days to come [12].

5.3.2 Social media, artificial intelligence information extraction, and situational awareness

The ability to notice things in one's surroundings, comprehend the significance of those things, and anticipate how those things will act in the near future is what we mean when we talk about having situational awareness (SA). Social media platforms such as Twitter and Facebook would make a significant contribution to the rapid dissemination of information in the event of a disaster, and could be essential to improving both SA and disaster management. Social media generates an enormous amount of data that is beyond the scope of manual examination.

When a crisis happens [13], victims who are directly and indirectly affected by the tragedy typically post vast quantities of data (such as images, text, audio, and video) to a variety of social media platforms (e.g., Facebook, Twitter, Instagram, and YouTube). These uploads can include anything from photographs to text to speech to video. This is owing to the fact that conventional avenues of communication for individuals to alert the public or emergency responders have been superseded by social media. Before reacting to an actual crisis, emergency personnel (also known as "first responders"), who come from a variety of emergency response organizations (EROs) often try to learn more about the situation. However, as soon as a catastrophe occurs, social media platforms are quickly overwhelmed with a variety of data. This flood of data is too much for emergency rooms that are equipped with big data to handle. Much of the data may be redundant or useless, and it becomes difficult for EROs to understand and base judgments on the large data that is available. Artificial intelligence plays a vital part in the processing and analysis of the huge volumes of social media data and in transforming it into knowledge that is logical and actionable.

Unstructured and unbalanced text streams can be found online [14]. A multi-label classification algorithm for the catastrophe text is originally built based on the historical dataset and will be used to develop a SA model from the generated huge, real-time, and noisy disaster text flow. This method assists in assigning the appropriate event tags to each disaster situation. A fresh approach to machine learning is created for the dynamic assessment of catastrophe risk for online text. Disaster events are regarded as featured vectors. The task of assessing the risk of a disaster is then changed to one that involves many classifications depending on the event, the users, and geographical concerns. This proposed model can provide disaster situation awareness as a result of online quantitative risk assessment findings based on actual accumulated precipitation data. These findings show that the proposed machine learning model can realize bottom-up automatic disaster information gathering by efficiently processing user-generated content created by victims.

5.4 PUBLIC ISSUES

When recommending the use of artificial intelligence (AI) in catastrophe risk communication, numerous issues must be considered.

Legal Concerns. Citizens make judgments during natural disasters based on the information provided by emergency agencies. Litigation alleging that the information provided to citizens was insufficient, inaccurate, or deceptive is not uncommon. Artificial intelligence-based systems in real-world operations might have unexpected repercussions, given that they learn and adapt autonomously and continually from their surroundings, facilitating or controlling the process of information creation and distribution. Managing and dealing with legal difficulties associated with AI usage in society, on the other hand, is a critical concern. The subject of autonomous machines' legal liability is still up for debate and has recently attracted a lot of interest in the literature [13, 15].

Quality of data. The adage "garbage in, garbage out" holds true in the context of machine learning algorithms. When the dataset is of poor quality, the algorithm will produce a sub-par result. Training AI systems can be compromised in a variety of ways: during data collection or cleaning, by introducing erroneous or incomplete data that does not accurately reflect the features of the population being studied, or simply by introducing the wrong data altogether. Before a dataset can be utilized in the process of machine learning, a significant amount of labor is thus typically required to clean it up. This may at times imply that incorrect data must be annotated with a broad variety of probable faults and how to remedy them. This is done so that the machine can detect similar problems in the future and manage them appropriately.

Controllability is also a big reason why AI isn't used more often in communicating about disaster risks. When there is an emergency, like a natural disaster, the authorities are expected to be in charge of the emergency operations. They should be able to decide what messages are sent, who gets them, and when they are sent. The use of chatbots and other AI-based communication systems could make it harder to keep such a tight grip on things. This is a problem that needs careful thought, because it is hard to take responsibility for something that cannot be changed. Trust is the other side of this problem. It's possible that emergency services won't be ready to fully deploy AI to communicate with the public about catastrophe threats until they can be certain that the system is always dependable, accurate, and powerful enough to handle the complex reality of emergency operations. Trust issues can also be caused by the actions or words of individual persons. If individuals who are in danger do not have faith that the automated communications they get from AI machines are accurate, it is possible that they may not take the appropriate precautions to protect themselves.

Influence of culture. It has been reported that culture has a substantial effect on risk perception [16]. Culture and language are interdependent; taking cultural differences, and the interdependence of language and culture, into account, an emergency warning may be understood differently in multicultural communities. Culture and ethnicity have not been adequately considered in the use of AI in disaster risk communication. Future research needs to focus on developing culturally intelligent AI robots that are able to generate communication on the dangers of natural catastrophes (for example, personalized emergency warnings), based on the ethical and communication protocols of the various linguistic and ethnic groups making up the community that has been studied.

5.5 ARTIFICIAL INTELLIGENCE IN DISASTER MANAGEMENT: A QUICK OVERVIEW

Using machine learning in disaster management may help remove irrelevant data, speed up data processing and analysis, and aid in all aspects of disaster response. It is not possible with ordinary machine learning (ML) methods to learn complex systems directly from raw data. Deep learning, a subcategory of ML, can automatically learn the representation of a complex system through the process of experimentation. This can be useful for achieving goals such as prediction, detection, or classification. It is possible to build increasingly complex and abstract models of the real system using neural network (NN) layers in deep learning (DL). Invariant qualities and highly complicated functions may be learned with DL techniques because they employ simple, non-linear modules to transform a representation into a higher, abstract one at each level [17]. Developments in DL have the potential to assist in the enhancement of disaster management. Because convolutional neural networks (CNNs) account for the majority of computer vision tasks, the use of satellite and aerial imaging systems is essential in the process of disaster response and damage assessment [18]. ANNs are becoming increasingly used as a potent tool for analyzing large datasets [19, 20]. These text-based NNs, such as LSTM and more recently Transformer, employ their architecture when it comes to natural language processing (NLP) [21]. NNs of this type are utilized in social media datasets for assessing the damage caused. The next step is to establish a theoretical foundation for disaster management by examining two common DL and ML architectures, CNN and LSTM, as well as SVM.

Convolutional neural networks: CNN architecture is made up of convolutional layers (CLs). In most cases, the application of the data in these layers is accomplished using table-based multiplication of n*m tabular filters, which are also referred to as kernels. Depending on the filter used, this procedure generates a variety of distinct representations of the incoming data. As the layers of the convolutional layer's filters operate, different features are revealed and projected in feature maps, which measure the stimuli they produce [17, 22]. After the convolution, there are many ways to change the input data. Kernel sizes and the number of kernels utilized in the convolution are critical to the network's overall efficiency. After the data has been convolutionalized, a pooling layer (PL) will gather the results of the convolution and will only save the most significant ones. These significant outcomes may include the maximum, minimum, or average of each value in the data. Before the input is allowed to be served into a completely connected layer for classification, it is first flattened in a one-dimensional space and then repeated several times, the exact number of which is determined by the depth of the network. The softmax layer performs a transformation on the data such that it becomes a probability distribution. It is important to keep in mind that the process of feature extraction receives an additional degree of abstraction from each CL that the network possesses.

Responses from social media include both pictures and words. CNNs may be utilized to extract both graphical and textual aspects from social media postings. The extracted visual and textual characteristics are joined together to produce a fused feature, which is then employed in the final phase of the classification process. This step occurs after the system has been adequately trained on the training sets. The outcomes of the CNN algorithm are effective in learning a variety of characteristics, including textual and visual. When just the textual characteristics are employed, the reliability of the system is worse than when the fusion feature is included, since it provides a visual feature. The postings that are relevant to the topic are automatically organized into categories according to the content and images that are contained within them. This is a documentation of a disaster that is taking place in real time during an event. When paired with the rich geographical settings that geotagging provides, social media might be of tremendous assistance in the prevention of disasters in many different ways.

Long short-term memory, often known as LSTM, is used for word classification in social media catastrophe datasets. LSTM circuits are examples of recurrent neural networks (RNNs). RNNs repeatedly perform the same algorithm on sequential data while passing along some information. In each time step, an input prediction affects subsequent predictions. This technique lets the network interpret complex textual material and derive meaning from word positions. LSTMs have a different cell design than RNNs. LSTMs use constant error carousels (CECs) to determine how much existing state information to pass along with incoming input data [22]. LSTMs' forecasts differ. The output of the network may be used to classify data using a probabilistic classifier, it can be used to forecast the next element in a series, or it can be used to predict a new sequence of elements. Learning rate and network size are crucial LSTM hyperparameters. LSTMs' independent hyperparameter tuning saves time during training and experimentation [23]. In every level of disaster management, accurate and exhaustive textual data analysis is essential, and LSTMs have shown themselves to be effective in this endeavor.

Support vector machine (SVM) is a basic yet powerful ML technique for classification and regression. It requires a labeled training set; however, support vector clustering may categorize unlabeled data [23]. SVM is straightforward. It produces a hyperplane that separates data while maximizing margin. A hyperplane in n-dimensional Euclidean space serves the function of partitioning the space. Margin is the smallest distance between the hyperplane and each category's nearest items. Maximizing the margin helps the algorithm identify groups, allowing for better predictions. Using a kernel method, SVM can do non-linear classification [24]. The data is augmented in a multidimensional space by using a kernel technique, which results in the characteristics being more clearly distinguishable from one another. The hyperplane is projected nonlinearly to the original plane. The kernel approach allows SVMs to learn complicated dataset invariants while remaining simple and quick. Many recent disaster management studies have started using SVM, which can provide remarkable results without sophisticated DL systems.

The effectiveness of disaster management activities may be assessed using AI-based methodologies. Monitoring, for example, helps keep tabs on the progress of emergency response and recovery efforts. For catastrophe monitoring, hybrid AI approaches are being employed [21, 25]. Disaster management operations should also be evaluated in terms of the suffering of the affected population [26]; AI-based methods analyze social media data to gage the suffering and feelings of the affected population during disaster response and recovery [27]. This will help disaster managers better deal with the complexity of operations. [26] Similarly, [27] these methods will help AI-based methods. Disaster operations necessitate the use of strong, tested, and trustworthy artificial intelligence systems. Human specialists should also be able to understand both the conclusion and the process that led to it in robust AI models.

5.6 CONCLUSION

The loss of human life as well as damage to infrastructure and property is one result of natural disasters. Artificial intelligence has made significant strides in recent years, and this has led to its increased usage in disaster management. To improve disaster recovery operations, future research should focus on using hybrid AI approaches to enhance mitigation efforts, remove vulnerabilities, and evaluate the resilience of vital infrastructure after a disaster has occurred. Because of the importance and complexity of catastrophe operations, it is necessary to use AI solutions that have been rigorously tested. Because disaster relief activities directly affect people's lives, the models used should be understood by experts and decision-makers on the ground [28]. Even more importantly, research should be devoted to creating innovative data-capturing methods and utilizing crowdsourcing to increase the efficacy of AI-based disaster management strategies.

AI can be used to save lives in a lot of different ways. One way is to help people deal with natural disasters. By combining the power of AI with a vision for a circular economy, we can take advantage of one of the most important technological advances of our time to an unprecedented degree. It will contribute to fundamentally changing the economy into one that is regenerative, resilient, and good for the long term.

REFERENCES

1. CRED CRUNCH from "EM-DAT: The OFDA/CRED International Disaster Database" Analysis & Writing by the EM-DAT Team in collaboration with UNDRR Centre for Research on the Epidemiology of Disasters (CRED), IRSS, UCLouvain.
2. Bourova, E.; Maldonado, E.; Leroy, J.B.; Alouani, R.; Eckert, N.; Bonnefoy-Demongeot, M.; Deschatres, M. A new web-based system to improve the monitoring of snow avalanche hazard in France. *Nat. Hazards Earth Syst. Sci.* 2016, 16 (5), 1205–1216.
3. Stoffel, A.; Brabec, B.; Stoeckli, U. GIS applications at the Swiss Federal Institute for Snow and Avalanche Research. In: *Proceedings of the 2001 ESRI International User Conference*, San Diego, 2001.
4. Asim, K.M.; Idris, A.; Iqbal, T.; Martínez-Álvarez, F. Earthquake prediction model using support vector regressor and hybrid neural networks. *PLoS One* 2018. https://doi.org/10.1371/journal.pone.0199004

5. Chang, M.-J.; Chang, H.-K.; Chen, Y.-C.; Lin, G.-F.; Chen, P.-A.; Lai, J.-S.; Tan, Y.-C. A support vector machine forecasting model for typhoon flood inundation mapping and early flood warning systems. *Water* 2018, 10, 1734. https://doi.org/10.3390/w10121734

6. Ahmad, K.; Riegler, M.; Pogorelov, K.; Conci, N.; Halvorsen, P.; De Natale, F. JORD: A system for collecting information and monitoring natural disasters by linking social media with satellite imagery. In: *Proceedings of the 15th international workshop on content-based multimedia indexing*, Florence, Italy. ACM, Article No. 12, 2017.

7. Arslan, M.; Roxin, A.; Cruz, C.; Ginhac, D. A review on applications of big data for disaster management. In *Proceedings of the 2017 13th International Conference on Signal-Image Technology & Internet-Based Systems (SITIS)*, Jaipur, India, 4–7 December 2017; pp. 370–375.

8. Van Wassenhove, L.N. Blackett memorial lecture humanitarian aid logistics: Supply chain management in high gear. *J. Oper. Res. Soc.* 2006, 57, 475–489. [CrossRef].

9. Altay, N.; Green, W.G. OR/MS research in disaster operations management. *Eur. J. Oper. Res.* 2006, 175, 475–493.

10. United Nations Office for Disaster Risk Reduction (UNDRR). UNISDR Terminology on Disaster Risk Reduction; UNISDR: Geneva, Switzerland, 2009. Available online: https://www.unisdr.org/files/7817_UNISDRTerminology English.pdf (accessed on 4 October 2021).

11. Ibarreche, J.; Aquino, R.; Edwards, R.M.; Rangel, V.; Pérez, I.; Martínez, M.; Castellanos, E.; Álvarez, E.; Jimenez, S.; Rentería, R.; Edwards, A.; Álvarez, O. Flash flood early warning system in Colima, Mexico. *Sens. J.* 2020, 20(18), 5231. https://doi.org/10.3390/s20185231

12. Asim, K.M., Martínez-Álvarez, F., Basit, A., Iqbal, T. Earthquake magnitude prediction in Hindukush region using machine learning techniques. *Nat. Hazards* 2017, 85, 471–486.

13. Broek, B..; Jakubiec, M. On the legal responsibility of autonomous machines. *Artif. Intell. Law* 2017, 25(3), 293–304.

14. Alam, F., Imran, M., Ofi, F. Image4Act: Online social media image processing for disaster response. In: *Proceedings of the 2017 IEEE/ACM international conference on advances in social networks analysis and mining 2017 (ASONAM'17)*. IEEE, pp. 601–604, 2017.

15 Hage, J. Theoretical foundations for the responsibility of autonomous agents. *Artif. Intell. Law* 2017, 25(3), 255–271.

16. Appleby-Arnold, S.; Brockdorff, N.; Jakovljev, I.; Zdravkovi, S. Applying cultural values to encourage disaster preparedness: Lessons from a low-hazard country. *Int. J. Disaster Risk Reduct.*, 2018, 31, 37–44.

17. Lecun, Y.; Bengio, Y.; Hinton, G. Deep learning. *Nature* 2015, 521, 436–444.

18. Presa-Reyes, M.; Chen, S.C. Assessing Building Damage by Learning the Deep Feature Correspondence of before and after Aerial Images. In Proceedings of the *2020 IEEE Conference on Multimedia Information Processing and Retrieval (MIPR)*, Shenzhen, China, 6–8 August 2020.

19. Yu, M.; Yang, C.; Li, Y. Big data in natural disaster management: A review. *Geosciences* 2018, 8, 165. [CrossRef]

20. Akshya, J.; Priyadarsini, P.L.K. A hybrid machine learning approach for classifying aerial images of flood-hit areas. In *Proceedings of the 2019 International Conference on Computational Intelligence in Data Science (ICCIDS)*, Chennai, India, 21–23 February 2019.

21. Fan, C.; Wu, F.; Mostafavi, A. A hybrid machine learning pipeline for automated mapping of events and locations from social media in disasters. *IEEE Access* 2020, 8, 10478–10490.

22. Schmidhuber, J. Deep Learning in neural networks: An overview. *Neural Netw.* 2015, 61, 85–117.

23. Ben-Hur, A.; Horn, D.; Siegelmann, H.T.; Vapnik, V. A support vector clustering method. In *Proceedings of the 15th International Conference on Pattern Recognition. ICPR-2000*, Barcelona, Spain, 3–7 September 2000.

24. Hofmann, T.; Schölkopf, B.; Smola, A.J. Kernel methods in machine learning. *Ann. Statist.* 2008, 36, 1171–1220.

25. Domala, J.; Dogra, M.; Masrani, V.; Fernandes, D.; D'Souza, K.; Fernandes, D.; Carvalho, T. Automated identification of disaster news for crisis management using machine learning and natural language processing. In *Proceedings of the 2020 International Conference on Electronics and Sustainable Communication Systems (ICESC)*, Coimbatore, India, 2–4 July 2020.

26. Drakaki, M.; Tzionas, P. Investigating the impact of site management on distress in refugee sites using Fuzzy Cognitive Maps. *Int. J. Disaster Risk Reduct.* 2021, 60, 102282.

27. Reynard, D.; Shirgaokar, M. Harnessing the power of machine learning: Can Twitter data be useful in guiding resource allocation decisions during a natural disaster? *Transp. Res. Part D Transp. Environ.* 2019, 77, 449–463.

28. Holzinger, A.; Dehmer, M.; Emmert-Streib, F.; Cucchiara, R.; Augenstein, I.; Del Ser, J.; Samek, W.; Jurisica, I.; Díaz-Rodríguez, N. Information fusion as an integrative cross-cutting enabler to achieve robust, explainable, and trustworthy medical artificial intelligence. *Inf. Fusion* 2022, 79, 263–278.

Chapter 6

IoT-based improved mechanical design for elevator cart

Reetu Malhotra and Armaan Jain

Chitkara University Institute of Engineering and Technology,
Chitkara University, Patiala, India

CONTENTS

6.1 INTRODUCTION

Modern elevators use a simple pulley-counterweight system. Although this system is reliable, power consumption by the motor and/or by the internal components of the cart is high, and motor temperature control is poor. This paper proposes a novel IoT-based system that uses the principles of mechanical advantage to increase payload, save power and sense the presence of people inside the elevator cart. Temperature and humidity control, improved cellular reception and a multi-pulley system are proposed, with the whole system connected to the cloud to enable data collection and monitoring of faulty parts. E-controls are an added convenience. The IoT functionality of this system was tested on a smaller scale using a prototype model and CAD simulations.

DOI: 10.1201/9781003415466-6

6.2 A CLOSER LOOK AT THE ELEVATOR

An elevator is a vehicle that moves vertically to transfer people and luggage. The shaft itself houses the operational machinery, motor, cables, and accessories. Human, animal, or water-wheel power was utilized to drive primitive elevators [2] dating from the third century BCE. Archimedes, a Greek scientist, was the first to construct an early form of what we now call an elevator. Even though researchers have come a long way from the raising platforms using pianos and cranes [3], the counterweight elevator concept remains the same as that designed by William Strutt (1756–1830).

According to data from the Center for Protection of Workers' Rights (CPWR), elevators are responsible for 60 percent of major injuries and 90 percent of deaths involving elevators and escalators [4, 21]. The accident rate is consistent across the world. Even in India, 28 deaths were directly attributed to elevators in 2019 [14, 15]. Suffocation due to high temperatures [22], defective alarm systems [26], and brake failure [20] are among the causes. Modern elevators pose a serious hazard to human life if not properly maintained, because the tension on the rotors and the rope itself wears the system down over time. This research proposes a variety of design enhancements to the existing elevator [25] that can significantly improve user experience through more robust temperature control [28], while also making it easier to use. This overhaul would enhance the cart's safety index from 60% [21, 27] to a possible 90%. Although the proposed new pulley system would necessitate a complete rebuild of the elevator, the IoT enhancements can be deployed with little to no disruption to the current system.

The remaining sections of this chapter cover IoT and devices, design and description of the new model, cellular reception, temperature modulation and user interface, conclusion and references.

6.3 IoT AND DEVICES

The Internet of Things (IoT) is a new technology prototype that aims to build a global network of machines and gadgets that can communicate with one another [7]. Its goal is to connect real-world and virtual-world things, allowing them to interact at any time and in any location. This results in a world where physical objects can be commanded and controlled by software built in the virtual world, allowing humans to use them more easily. Many real-world IoT applications have been presented in a variety of industries, and this dominion extends beyond traditional industrial areas to everyday life, where IoT can provide major improvements and even lead to new business models [5, 6].

6.4 THE TECH BEHIND AN ELEVATOR

For security reasons, people used to avoid traveling in remote locations or late at night. Gazdzinski [11] described a secure personnel transport system and elevator for modern urban life. This control and information system enhanced security in existing elevator devices. Yost et al. [12] described the work of an elevator system configuration engineer, including the information the customer must provide at the outset, how to use that information to put together a safe and effective elevator system, and what information must be provided to the installers and inspectors at the end [10].

Traditionally, elevators have been high energy consumers, in particular where there was a mismatch between the elevator system scheduler and the moving carts. Van et al. [14] developed an energy-saving elevator with IoT communications which reduces unnecessary cart moves. Studying how the round-trip time was distributed was beneficial to the performance analysis of elevators during peak hours.

Jiang et al. [6] described IoT as an essential vehicle necessary for real business applications, and examined elevator manufacturers' moves towards a more IoT-based approach to the system. The authors also discussed the implications for the industrial system and explained the differences between S-D logic and G-D. The latest IoT technology has enabled companies to improve skills and promote information sharing. Lee et al. [7] explain the importance of five technologies for the exploitation of successful IoT-based commodities/services, and how IoT can increase customer value. Schiff et al. [15] examined and assessed strategies for adding damping to improve system responses, as well as the impacts of rope dynamics on system responses. Tarmo et al. [16] used deep learning to create passenger profiles and investigated the behavior and travel requirements of elevator passengers, as well as the feasibility of predicting passenger destination floor. Their groundbreaking research, which was conducted using an actual smart elevator system installed in a typical office building, aided in the establishment of the monitoring mechanism that IoT devices would eventually rely on. Lee et al. [17] discussed the group elevator effect in their study. In the earlier studies, some cages were taken idle. The cages were assumed to go back to the first floor, which resulted in loss of energy [34]. The authors [18–19] developed an idle cage assignment algorithm to tranship the idle cages. Papers [20–31] highlight the accidents, injuries and deaths that have occurred due to the faulty design of the current system, elevator safety checks, and wireless multimedia sensor networks topologies. Several researchers [35–42] studied how to minimize elevator costs, including installation costs, preventative maintenance costs, design costs etc. Different algorithms such as genetic algorithms [43–47], searching algorithms, and reliability models [48–50] are used to optimize the costs and timings of systems.

6.5 DESIGN OF THE NEW MODEL

We consulted a number of important papers for the design of our model:

i. Nick Health [8] helped us understand the advantages of using the Raspberry Pi.
ii. Robert [11] helped with the modeling of the new panels.
iii. [12] includes points to take into consideration in configuring the elevator and its safety measures.
iv. [13] helps with the algorithms used in the analysis of both old and new elevator designs.
v. [6] and [7] highlight the growing adoption of IoT culture in the elevator business.
vi. [16] helps with creation of the IoT-based monitoring system.

The proposed model (Figure 6.1) uses an integrated pulley system to **increase the effective payload** of the elevator on existing motor torque, using the weight of the cart itself as the counter and thus eliminating the counterweight altogether.

A number of pulleys are used at either end of the elevator housing space in order to reduce the amount of force necessary to move an object by increasing the amount of rope used to raise it. The length of rope can be determined by using the formula:

New length of rope = Original length of the rope
                     × (the number of additional pulleys/2)

The chances of breakage and the wear and tear on the cart's cable are thus considerably reduced. The length of the rope is doubled with every two

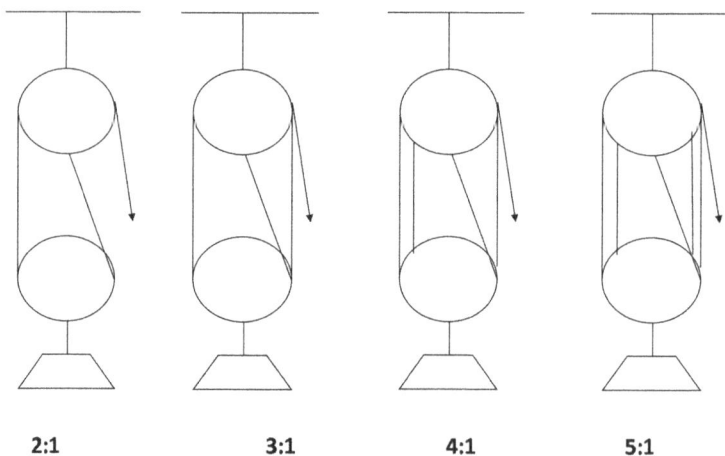

2:1   3:1   4:1   5:1

Figure 6.1

additional pulleys, but for every two additional pulleys the effective weight of the cart is reduced by 50%.

A Raspberry Pi 3 b [8] is used in the elevator to achieve the new and enhanced temperature modulation, cellular reception, and cloud connection. It acts as the main processor and provides cloud connectivity. Other devices are listed below and detailed in the sections that follow:

1. Distributed Antenna System (DAS) for the cellular reception
2. DHT22 temperature and humidity sensing
3. Fans for temperature control
4. Ultrasonic sensors to detect number of passengers in the cart [52], [51]
5. (Alternative to 4) Array of pressure sensors underneath the cartpet
6. Mesh network-based smart lights for power management
7. Interconnected touch input for synchronizing with E-inputs
8. Microphone for voice input.

The building information modeling (BIM) is implemented by Isihdag [9]. His paper "BIM and IoT: A Synopsis from GIS Perspective" is a good starting point for the proposed cloud computing model [11–12].

The IoT model will also collect data on the frequency of usage and alert the respective authority for timely maintenance. The same setup can also be used as an input for detection of any faulty devices (1–5).

6.5.1 Temperature modulation and user interface

When the cart is empty, it uses devices 3–6 to shut down systems such as lights and fans, greatly reducing power consumption. The temperature sensor (device 2) monitors the temperature of the cart every quarter of an hour. If the temperature increases from the set temperature, the fans are triggered. Alternatively, if requested by the customer, heaters can be installed in the cart and activated if the temperature falls below the set point.

As soon as the cart door opens and a user is sensed as about to enter, devices 3–6 trigger the lights, device 2 is used to sense the temperature again, and a suitable action is performed. A default is set for every smart device and any setting can easily be changed through the E-control panel.

- For the convenience of the user, an improved control panel is proposed in this work. It includes a cancel feature, pre-set temperature display screen and buttons for pre-set temperature modulation. The cancel feature allows the user to double-click the floor number in order to prevent other floors' buttons being accidentally depressed.
- Pre-set temperature display displays the current temperature that is being maintained in the cart.
- Temperature modulation buttons, used to adjust the pre-set temp.

- E-calling saves time by calling the elevator beforehand from any part of the building.
- "Hold the Door" voice command holds the door open to prevent users just missing the elevator.

The panel is equipped with NFC trigger tags in case of failure of the physical input module. These NFC triggers are also to be added to the outer physical inputs for hygiene reasons.

6.6 CELLPHONE RECEPTION

Metal elevators have poor connectivity as wireless signals cannot penetrate through thick materials. This has implications for both convenience and public safety.

None of the models currently in use provides the desired smooth and seamless coverage. The next section explores the limitations of some of these models.

6.6.1 Current in-cart cellphone coverage models in elevators

1. Antennas mounted **in the elevator** chamber at a certain height from the ground are not effective since they cannot provide consistent coverage throughout the length of the chamber in tall buildings. Nor can repeaters within the elevator shaft provide adequate signal strength (especially in taller buildings) to cover the entire shaft.
2. Multiple **network extenders** mounted in the elevator lobby after a certain number of floors cause a constant that degrades call performance and leads to call drops as a device changes from one coupler to another with the movement of the cart along the lobby.
3. Using an **omni-directional powerful antenna in the midst** of the elevator lobby is far more effective than all the other methods but has its own drawbacks. The antennas are expensive, prone to damage and are difficult to place. Additional wiring is also required.
4. Other systems like installation of **repeaters on each floor** and placement of a **high-power distributed antenna** at top and bottom have been tried and have had similar results.

6.6.2 Proposed solution to the problem of network coverage

The proposed solution to the problem of cellphone coverage is to use fiber-optic cables as antenna bands running along the length of the entire chamber on the inside. This obviates the need for running heavy network cables alongside the power cables for the elevator cart.

Figure 6.2

Figure 6.3

Figures 6.2–6.4 show, respectively, the diagonal, front and top views of the shaft structure without the bands installed. Figures 6.5–6.9 show the front, diagonal, inside, top and see-through views of the shaft structure with the bands (represented by blue lines) installed.

Figure 6.10 shows diagonal and inner-left view of the shaft with blue lines representing the bands. In this system, a single repeater/router placed on the cart itself will have a continuous supply of bandwidth across the entire

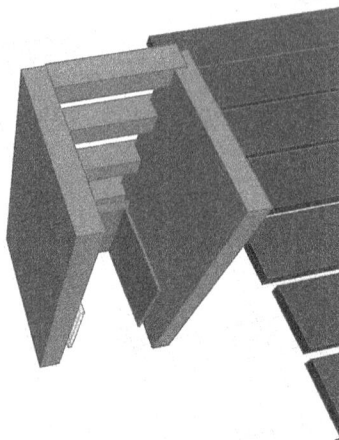

Figure 6.4

Figure 6.5

Figure 6.6

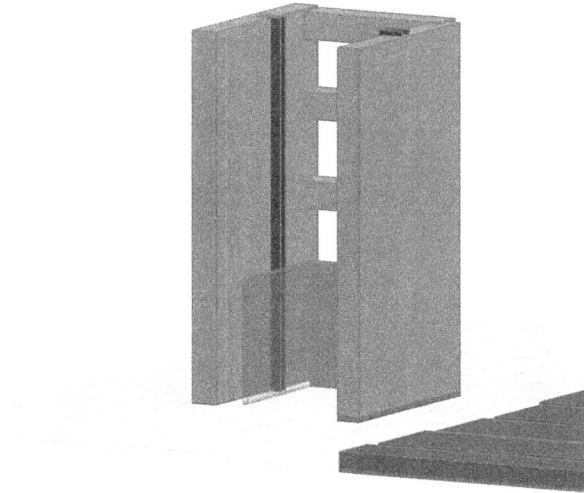

Figure 6.7

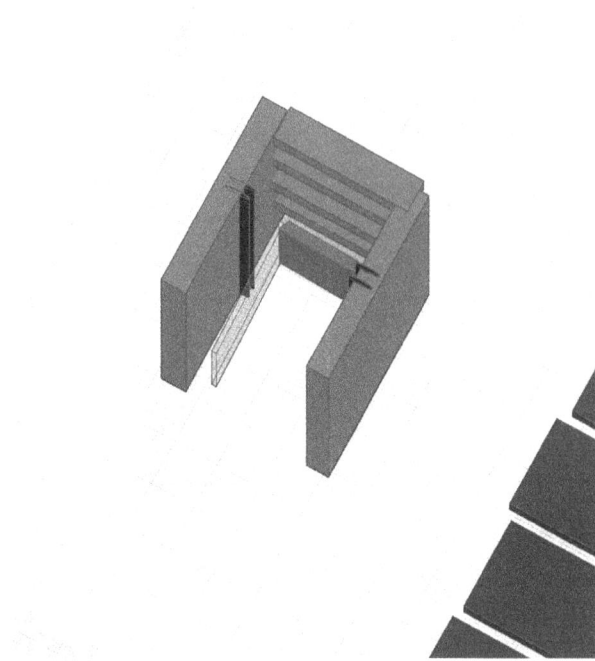

Figure 6.8

Figure 6.9

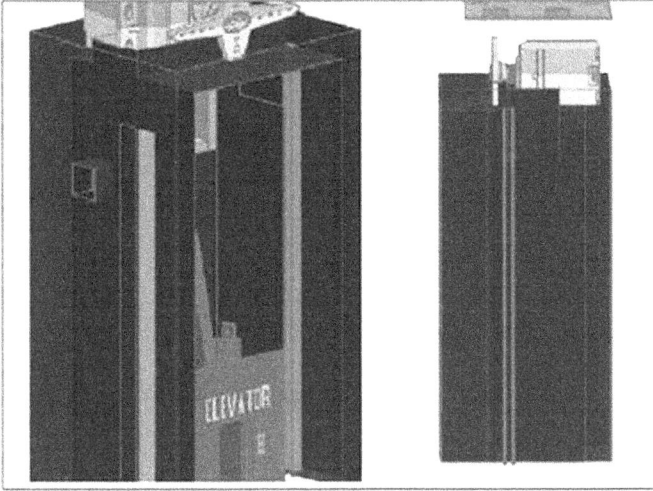

Figure 6.10

chamber. We can use coaxial cable instead of fiber-optic cable but that would be an expensive option. We can achieve continuous coverage between floors without any signal drop, and although the service zone is limited to the elevator cart itself and its immediate surroundings, service is continuous and smooth. The attenuation that could have been caused by other methods is thus eliminated. As Figure 6.10 shows, these bands do not cause any obstruction.

Figure 6.11 shows the top and diagonal views of the elevator cart, demonstrating the multi-pulley system and the router placement. Figure 6.12 is a functional representation of the whole lift.

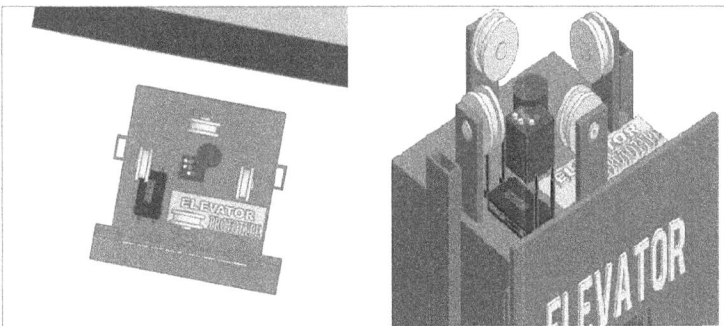

Figure 6.11

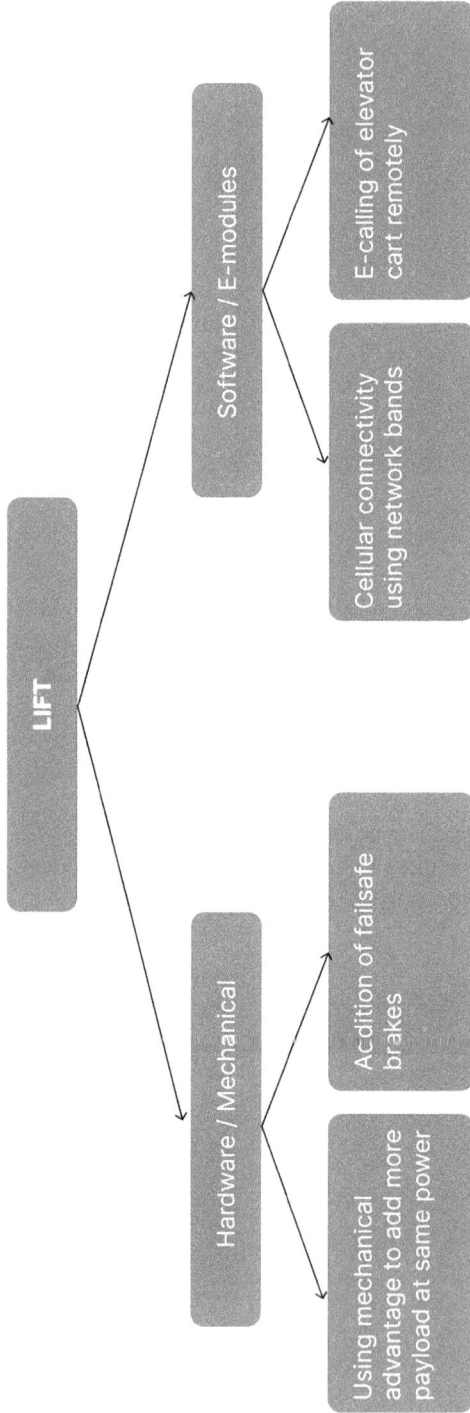

Figure 6.12

6.7 RESULTS AND FINDINGS

The new model aims to increase the cart's safety index by 20–30% by combining the new pulley system with the old design's safety measures. With the new temperature and ultrasonic sensors, the cart becomes significantly more energy efficient, reducing electricity use by 50% as a starting point. Energy-saving data would be regularly collected and could be utilized to fine-tune the system.

6.8 CONCLUSION

This study presents a novel approach to the modern elevator in which IoT is used in combination with physics. The new four-pulley arrangement uses the principles of mechanical advantage to not only provide better security for the users but also to help increase the longevity of the system as a whole. IoT-based modifications provide a better user experience with added temperature control and address the problem of excess energy consumption. The updates to the system help maintain system health and facilitate fault detection. Future improvements will include incorporation of 5G technology in the currently 4G-capable routers. As sensors gain precision with time, the size and cost of implementation will become more affordable.

ACKNOWLEDGMENT

There is no conflict of interest.

REFERENCES

[1] S. Abe and E. Watanabe, "History of elevators and related research", *IEEE Transactions*, vol. 124, no. 8, pp. 679–687, 2014.

[2] M. Bellis, "The history of Elevators from top to bottom", Thoughts Co article, August 11, 2019.

[3] L. E. Grey, *From Raising rooms to Express Elevators: A history of Passenger elevator in the 19th century*, ISBN 1-886536-46-5.

[4] CPWR, "Deaths and Injuries Involving Elevators and Escalators - A Report of the Centre To Protect Workers' Rights", July 2006.

[5] C. T. A. Lai, P. R. Jackson, and W. Jiang, "Internet of things business models in ecosystem context-cases of elevator services", *IJCSE*, vol. 3, pp. 1–12, 2018.

[6] W. Jiang, P. R. Jackson, and C. T. Lai, "Shifting paradigm to service-dominant logic via internet-of-things with applications in the elevators industry", *Journal of Management Analytics*, vol. 4, pp. 35–54, 2016.

[7] I. Lee and K. Lee, "The internet of things (IoT): Applications, investments, and challenges for enterprises", *Business Horizons*, vol. 58, pp. 431–440, 2015.

[8] Nick heath, "What is raspberry pi 3?" Znet.com article, (Nov. 30, 2017).

[9] U. Isihdag, "BIM and IoT: A Synopsis from GIS Perspective", Volume: XL-2/W4, 2015.

[10] G. D. Smart, "Method for installing elevator system", 1982.

[11] R. F. Gazdzinski, "Smart elevator system and methods", 2010.

[12] G. R. Yost and T. R. Rothenfluh, "Configuring Elevator system", 1996.

[13] Y. Lee, T. S. Kim, H.-S. Cho, D. K. Sung, and B. D. Choi, "Performance analysis of elevator during peak-up", *Mathematical and Computing Model*, 2009.

[14] L. Van, Y. Lin, T. Wu, and T. Chao, "Green Elevator Scheduling Based on IoT Communications," *IEEE Access*, vol. 8, pp. 38404–38415, 2020,

[15] A. J. Schiff, H. S. Tzou, and Y. H. Chu, "Earthquake response of elevator counterweights", 1980.

[16] T. Robal, U. Reinsaul, and M. Leier, "Towards personalised elevator travel with smart elevator system", *Baltic Journal of Modern Computing*, vol. 8, no. 4, pp. 675–697, 2020.

[17] S. Lee and H. Bahn, "An energy-aware elevator group control system," in *Proc. 3rd IEEE Int. Conf. Ind. Informat. (INDIN)*, Perth, WA, Australia, pp. 639–643, 2005.

[18] T. Zhang, S. Mabu, L. Yu, J. Zhou, X. Zhang, and K. Hirasawa, "Energy saving elevator group supervisory control system with idle cage assignment using genetic network programming", in Proc. IEEE ICCAS-SICE, pp. 994–999, 2009.

[19] J.-L. Zhang, J. Tang, Q. Zong, and J.-F. Li, "Energy-saving scheduling strategy for elevator group control system based on ant colony optimization," in *Proc. IEEE Youth Conf. Inf., Comput. Telecommun*, pp. 37–40, Nov. 2010.

[20] Z. Hu, Y. Liu, Q. Su, and J. Huo, "A multi-objective genetic algorithm designed for energy saving of the elevator system with complete information," in *Proc. IEEE Int. Energy Conf.*, pp. 126–130, Dec. 2010.

[21] J.-B. Park and S.-J. Shin, "Propose of elevator system for effective management based on IOT", *The Journal of The Institute of Internet, Broadcasting and Communication*, vol. 14, no. 6, pp. 163–167, Dec. 2014.

[22] T. Tyni and J. Ylinen, "Evolutionary bi-objective optimisation in the elevator cart routing problem", *European Journal of Operational Research*, vol. 169, pp. 960–977, 2006.

[23] N. Kobayashi, K. Kawase, S. Sato, and T. Nakagawa, "Experimental, verification of effects in an emergency stop by installation of magneto rheological fluid damper to an elevator", *IEEE Transactions on Magnetics*, vol. 53, pp. 1–4, 2017.

[24] Mamatha Sandhu, Reetu Malhotra, Jaspreet Singh. IoT Enabled -Cloud based Smart Parking System for 5G Service 2022 *IEEE Xlore*, 1st IEEE International Conference on Industrial Electronics: Developments & Applications (ICIDeA) | 978-1-6654-2149-2/22 ©2022 IEEE | DOI: 10.1109/ICIDEA53933.2022.9970012. pp. 202–207. International Conference on Industrial Electronics: Developments & Applications (ICIDeA). IEEE Kolkata Section Industrial Electronics Society Chapter- Bhubaneswar and IEEE Bhubaneswar Subsection on 15th – 16rh Oct 2022.

[25] M. Zhang and C. Huang, "Design of elevator real-time monitoring system based on internet of things", *The Chinese Test*, vol. 1, pp. 101–105, 2012.

[26] Y. Zhou, K. Wang, and H. Liu "An elevator monitoring system based on the internet of things", *Procedia Computer Science*, vol. 131, pp. 541–544, 2018.

[27] X. Yan, L. Xie, Z. Cheng et al. ZigBee +3G "The application of the network in the new type of well - road elevator monitoring system", *Automatic Instrument*, vol. 1, pp. 1–4, 2015.

[28] T. Melodia and I. F. Akyildiz, *Research Challenges For Wireless Multimedia Sensor Networks /Distributed Video Sensor Networks*, London: Springer, pp. 233–246, 2011.

[29] R. Malhotra, T. Dureja, and A. Goyal, "Reliability analysis a two-unit cold redundant system working in a pharmaceutical agency with preventive maintenance", *Journal of Physics: Conference Series*, vol. 1850 012087, no. pp. 1–10, 2021.

[30] H. Yan. "Research on the application of the internet of things in the elevator industry". *Industrial Control Computer*, vol. 8, pp. 102–103, 2013.

[31] D. Tao, *Research and practice of remote monitoring system for elevator Internet of Things*, Xia Men: Xia Men University, 2013.

[32] R. Malhotra, D. Kumar, and D. P. Gupta, "An android application for campus information system", *Procedia Computer Science*, vol. 172, pp. 863–868, 2020.

[33] D. Kumar, R. Malhotra, and S. Ram, "Design and construction of a smart wheelchair, Elsevier", *Procedia Computer Science*, vol. 172, pp. 302–307, 2020.

[34] J. R. Fernández, P. Cortés, J. Guadix, and J. Muñuzuri. "Dynamic fuzzy logic elevator group control system for energy optimization," *International Journal of Information Technology & Decision Making*, vol. 12, no. 3, pp. 591–617, 2013.

[35] M. Z. Hasan, R. Fink, M. R. Suyambu, and M. K. Baskaran Presented their paper "Assessment and improvement of intelligent controllers for elevator energy efficiency" at the IEEE International Conference on Electro Information Technology 2012. The conference was held in Indianapolis, USA in 2012.

[36] J. Liu and Y. Liu presented their paper "Ant colony algorithm and fuzzy neural network based intelligent dispatching algorithm of an elevator group control system" at the IEEE International Conference on Control and Automation. The conference was held in Guangzhou, China in 2007.

[37] J. Jamaludin, N. Rahim, and W. Hew Presented their paper "Self-tuning fuzzy-based dispatching strategy for elevator group control systems" at the Proceedings ICCAS. The conference was held in Seoul, South Korea in 2008.

[38] J. Jamaludin, N. Rahim, and W. Hew. "An elevator group control system with a self-tuning fuzzy logic group controller". *IEEE Transactions on Industrial Electronics*, vol. 57, no. 12, pp. 4188–4198, 2010.

[39] J. Fernandez, P. Cortes, J. Munuzuri, and J. Guadix. "Dynamic fuzzy logic elevator group control system with relative waiting time consideration", *IEEE Transactions on Industrial Electronics*, vol. 61, no. 9, pp. 4912–4919, 2014.

[40] C. Kim, K. Seong, and H. Lee-Kwang. "Design and implementation of a fuzzy elevator group control system," *Proceedings of the IEEE Transactions on systems man and Cybernetics*, vol. 28, no. 3, pp. 277–287, 1998.

[41] T. Ishikawa, A. Miyauchi, and M. Kaneko. "Supervisory control for elevator group by using fuzzy expert system which also addresses traveling time",

Proceedings of the IEEE International Conference on Industrial Technology, vol. 2, no, 1, pp. 87–94, 2000.

[42] N. Imasaki, S. Kubo, S. Nakai, T. Yoshitsugu, K. Jun-Ichi, and T. Endo Presented their paper "Elevator group control system tuned by a fuzzy neural network applied method" at the IEEE International Conference on Fuzzy Systems 1995. *International Joint Conference of the Fourth IEEE International Conference on Fuzzy Systems and The Second International Fuzzy Engineering Symposium.* The conference was held in Yokohama, Japan in 1995.

[43] M. M. Rashid, B. Kasemi, A. Faruq, and A. Z. Alam Presented their paper "Design of fuzzy based controller for modern elevator group with floor priority constraints" at the 4th International Conference on Mechatronics: Integrated Engineering for Industrial and Societal Development, ICOM'11. The conference was held in Kuala Lumpur, Malaysia in 2011.

[44] J. Alander, T. Tyni, and J. Ylinen Presented their paper "Elevator group control using distributed genetic algorithm" at the Proceedings of the international conference on artificial neural nets and genetic algorithms (ICANNGA95). The conference was held in Alès, France in 1995.

[45] A. Miravete. "Genetics and intense vertical traffic", *Elevator World*, vol. 47, no. 7, pp. 118–120, 1999.

[46] T. Chen, Y. Hsu, and Y. Huang. "Optimizing the intelligent elevator group control system by using genetic algorithm", *Advanced Science Letters*, vol. 9, pp. 957–962. 2012

[47] H. Kitano, Genetic Algorithms, Sangyo Tosho, Japan 328, 1993.

[48] J. Zhang, X. Zhao, C. Liu, and W. Wang, "Reliability evolution of elevators based on the rough set and the improved TOPSIS method" *Mathematical Problems in Engineering*, vol. 2018, pp. 1–8, 2018.

[49] S. Batra and R. Malhotra, "Reliability and availability analysis of a standby system of pcb manufacturing unit," *Reliability and Risk Modeling of Engineering System.* Chapter 6 EAI/Springer Innovations in Communication and Computing pp. 75–87, 2021.

[50] R. Malhotra, "Reliability evaluations with variation in demand", *Systems Reliability Engineering: Modeling and Performance Improvement*, edited by Amit Kumar and Mangey Ram, Berlin, Boston: De Gruyter, pp. 89–100, 2021.

[51] G. Taneja and R. Malhotra. "Cost-benefit analysis of a single unit system with scheduled maintenance and variation in demand", *Journal of Mathematics and Statistics, USA*, vol. 9, no. 3, pp. 155–160, 2013. doi:10.3844/jmssp.2013. 155.160.

[52] R. Malhotra, A. Goyal, M. Batra, and R. Bansal. "Preprogrammed multiway workbench- an experimental study using ultrasonic sensors with the help of microcontroller", *Journal of Physics: Conference Series*, vol. 1850 012086, pp. 1–7, 2021.

Chapter 7

Wearable IoT using MBANs

Khushboo Dadhich and Devika Kataria

JK Lakshmipat University, Jaipur, India

CONTENTS

7.1 INTRODUCTION: BACKGROUND AND DRIVING FORCES

The Internet of Things is a ubiquitous network and has attracted attention for real-time monitoring in the healthcare domain [1–7]. The use of IoT in healthcare is increasing, owing to the scarcity of healthcare professionals. Since just 3,71,000 of India's approximately 12,50,000 doctors are specialists, the availability of medical professionals has long been a cause for concern [8, 9]. The majority of medical professionals reside in cities and work long hours at many hospitals. A system is needed in which technology can supply healthcare to a vast rural population while also making specialists available remotely. Additionally, critical care patients require 24/7 supervision and monitoring even during night hours [10, 11]. Thus real-time monitoring of patients' health indicators, with secure transmission of this information to specialists is necessary for patients in intensive care units. At present bedside monitoring generally involves wired sensors and probes which cause discomfort for the patient and restrict their mobility. Wireless sensors which are more comfortable to use for extended periods of time

DOI: 10.1201/9781003415466-7

have become more popular because of advances in sensor technology [12, 13]. Flexible wireless sensor systems requiring a small amount of processing power are easily accessible and consume less energy.

7.2 THE IEEE 802.15.6 STANDARD

MBANs use connected sensor nodes to detect and send signals to a body network hub (BNH), which then transmits the information to a central server [14]. MBAN sensors are either applied to the skin or embedded within the body. These sensors allow the patient to be mobile within the health care center while being monitored for health parameters (Figure 7.1). The nodes send data to the BNH in a sequence as decided by the priority assigned to them (refer Table 7.1), and the BNH forwards the data to a medical server in real-time. Various operational aspects and design standards for personal health information (PHI) are to be considered for such applications. The data from the sensors is transmitted securely, and data access authentication must be established at the server [14, 15].

MBANs are resource constrained in terms of energy consumption, computational power, communication range, and bandwidth. The electromagnetic radiation from the sensors must not harm the human body when the sensors are near the skin. Power consumption by the sensors must be low and the lifetime of the batteries should be long enough to avoid frequent

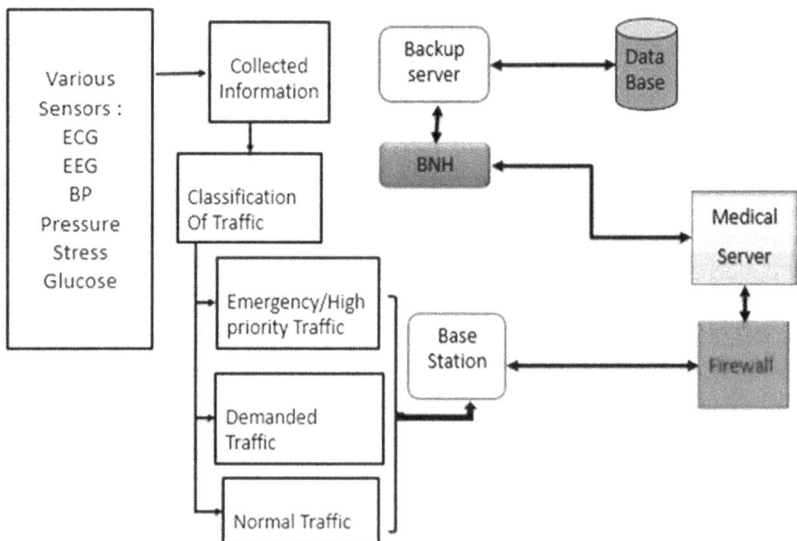

Figure 7.1 Medical body area network.

Table 7.1 Types of traffic on the bases of priorities

Type of traffic	Priorities of user device
Emergency event report	UP 7
High-priority data	UP 6
Medical data	UP 5
Voice	UP 4
Video	UP 3
Excellent effort	UP 2
Best effort	UP 1
Background	UP 0

replacements. The data transmission must be reliable and priority allocation is essential for some parameters to be measured. In 2012, IEEE standard 802.15.6 was developed to allow limited-power, short-range, highly reliable, and secure wireless systems for MBANs, with wide applications for personal health monitoring.

7.3 OVERVIEW OF THE 802.15.6 STANDARD

The IEEE 802.15.6 standard has two layers: physical and medium access control (MAC). The physical layer has three sublayers: the human body communication layer (HBC PHY), which runs between 5 and 50 MHz, the ultra-wideband layer (UWB PHY), which operates between 3100 and 10600 MHz, and the narrow band layer (NB PHY), which operates between 402 and 2450 MHz [16, 17]. According to Table 7.2, which contrasts IEEE 802.15.6 with IEEE 802.15.4, medical nodes of MBAN can communicate over a shorter distance since they consume less power, however, they can carry data at a higher rate [18].

This standard offers a single-star or two-hop extended star architecture to increase the delivery via different paths, as shown in Figure 7.2. The nodes in a single-hop star topology are linked to a single BNH; in two-hop technology, an intermediate or relay node is used to facilitate communication between the BNH and the nodes [14–16].

A single-hop star topology is suitable when the network is stable and the connection between BNH and nodes has been consistently established for a long time. However, because of the longer distance for data travel and limited power available with nodes, many times two-hop extended topology is used. The handshake procedure for the two-hop topology, which uses a relay node to exchange data frames from a node to a BNH, is shown in Figure 7.3. Relayed nodes (sensor nodes) that are unable to access the BNH

Table 7.2 IEEE standards 802.15.4 and 802.15.6 compared

Parameters	IEEE standard	
	802.15.4	*802.15.6*
Types of frames used in MAC	Beacon, command, response, and data	Management, control, and data
Mechanism	Slotted CSMA/CA, Unslotted CSMA/CA	CSMA/CA, Slotted Aloha, improvised and unscheduled access, scheduled and scheduled polling access
Data packet Rate (Max)	20 kbps, 40 kbps, 250 kbps	15.6 Mbps
Network Range	75 m	3 to 5 m
Applications	Lower data transfer rate Sensors for commercial use Smart grid	Portable sensors

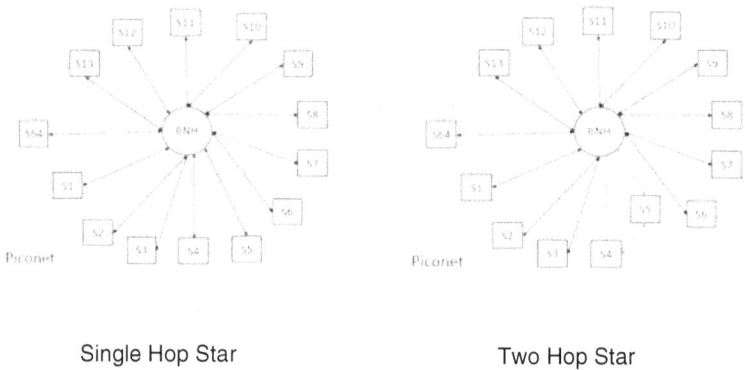

Single Hop Star   Two Hop Star

Figure 7.2 Piconet of MBANs.

immediately contact the relaying node by submitting a request to establish a connection. As a response to the connection setup request, the relaying node sends an acknowledgment frame and forwards the request to BNH. Following receipt of the acknowledgment frame, the BNH offers a slot for the relayed node to the relaying node and BNH interface. The relayed node then delivers the data frame to the relaying node, which is forwarded to the BNH and receives a response from the BNH in the form of an acknowledgment frame which is forwarded to the relayed node. [19–21].

The BNH acts as a centralized controller, allocating channels to MBAN nodes and communicating with nodes using three different access modes as discussed in the next section.

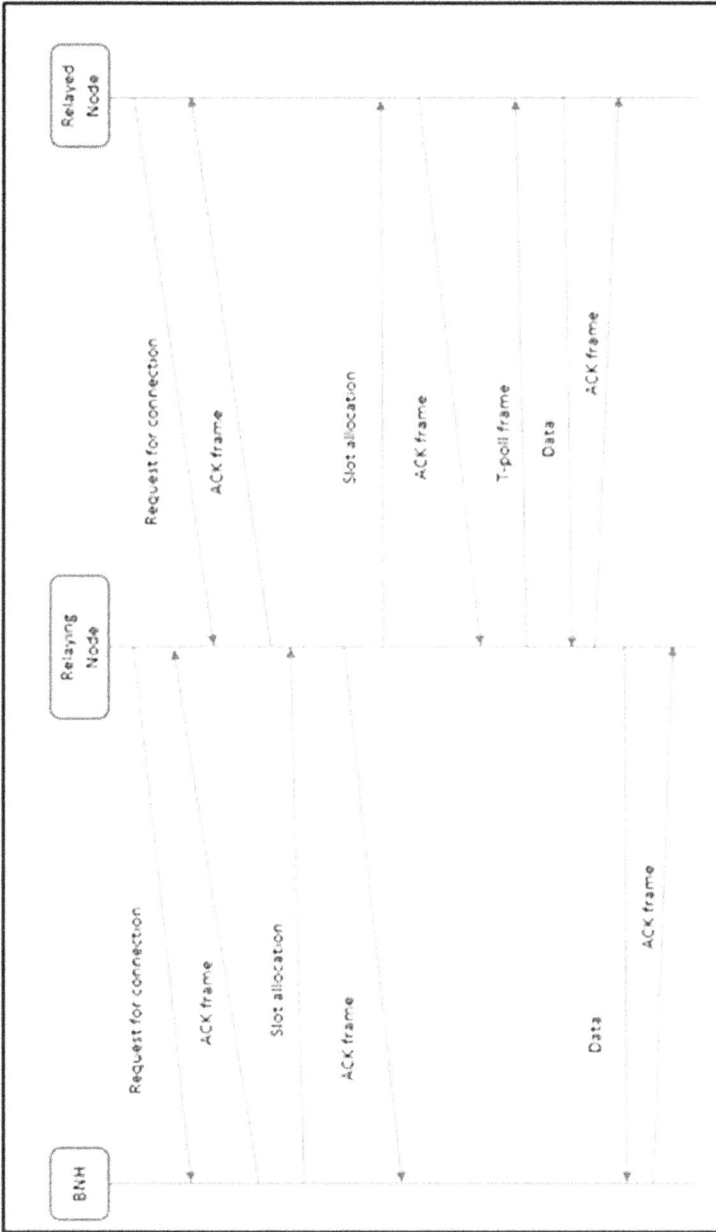

Figure 7.3 Relaying protocol process.

7.4 CHANNEL COMMUNICATION MODES FOR MBAN

The IEEE 802.15.6 MAC layer allows one of the following three modes for synchronization of data communication between the BNH and the MBAN nodes:

1. Beacon mode with beacon period or super-frame boundaries: Several different time intervals inside a single super-frame are assigned to the nodes in this method for contention-free data delivery.
2. Non-beacon mode with super-frame boundaries: Used for uplink, downlink, and bi-link connections; only operates during the management access phase.
3. Non-beacon mode without super-frame boundaries: Sometimes higher-priority nodes are to be polled or posted allocations and this is done in the non-beacon mode without a super-frame.

The standard divides the channel access time into super-frames. The super-frame where a node is transmitting data is called an active super-frame and may be followed by non-active super-frames where there is no transmission.

In the beacon mode, BNH broadcasts a beacon frame B at the beginning and end of the super-frame, so as to synchronize with the nodes. To establish a connection with BNH the node sends a request mentioning the number of time frames required. A notification frame for the new connection is set in the MAC and scheduled by the BNH, which informs the node about the connection. During data transmission, MBAN nodes are assigned priority, as shown in Table 7.1. The MBAN nodes are assigned user priorities (UP) in the range of 0–7. UP6 and UP7 are used when any emergency data is to be transmitted by any node.

The super-frame is divided into various access phases: exclusive access phase (EAP1 and EAP2), random access phase (RAP1 and RAP2), managed access phase (MAP1 and MAP2), and contention access phase (CAP) as shown in Figure 7.4.

The node with the highest priority (UP7) can access EAP, while all other nodes may access the channel during the RAP. The MAP handles both planned and unplanned bi-link allocations, as well as planned uplink and

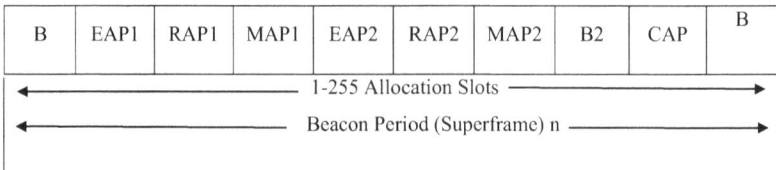

B	EAP1	RAP1	MAP1	EAP2	RAP2	MAP2	B2	CAP	B

1-255 Allocation Slots

Beacon Period (Superframe) n

Figure 7.4 Beacon mode with super-frame.

Table 7.3 Accessing phase and accessing methods

Accessing phase	Full-form	Description	Accessing methods
EAP1	Exclusive access phase 1	The highest priorities are data transmission or emergency data transmission	Node contention for resource allotment by CSMA/CA or Slotted Aloha
EAP2	Exclusive access phase 2		
RAP1	Random access phase 1	All priorities data transmission	
RAP2	Random access phase 2		
CAP	Contention access phase		
MAP1	Managed access phase 1	Handle uplink, downlink, bi-link, and delay bi-link allocation	Type I and II polling is used
MAP2	Managed access phase 2		

downlink allocations, polled and posted allocations. Each node must create a connection with the BNH to get a planned allocation period. During data transmission, if none of the nodes have emergency data, then the length of EAP is made zero and these time slots are used for RAP. When assigning channel slots to the nodes, the BNH can adjust the access phase lengths as zero, except for RAP1. The access phases and the accessing mechanism are listed in Table 7.3.

The allocations in EAP, RAP, and CAP are done through by CSMA/CA or Slotted Aloha. To transmit high-priority data, the BNH may merge the exclusive and random-access phases into a single EAP, as well as regard the RAP next to the EAP as EAP.

The MAC layer of IEEE 802.15.6 is crucial to maximizing network life and ensuring excellent service. At the MAC layer, several access mechanisms such as contention access (CA), polling access (PA), and scheduled access (SA) can be employed, depending on the application's needs. Uplink and downlink access strategies as well as an unscheduled polling/posting-based access strategies are utilized in the contention-free access phase, in contrast to the (CSMA/CA) carrier sense multiple access/collision avoidance and Slotted Aloha approaches that are used in the contention access phase [22, 23].

7.5 RESOURCE ALLOCATION

The resource allocation technique described in IEEE 802.15.6 starts with the nodes reporting quality of service (QoS) parameters like power consumption, data rate, probability of packet loss and latency. To establish a connection, a node sends a frame for accessing the slot to the BNH. After obtaining the access request frame, the BNH allocates timeslots and communicates through the connection assignment frame. Once connections have been set up, every node sends frames of information to the BNH with the set output. The data transmission may be done using CSMA/CA or Slotted Aloha protocol.

7.5.1 Slotted Aloha access

The Slotted Aloha access technique allows network access based on user priorities. When a node joins the transmission or retransmission allocation queue, a collision probability (CP) is assigned based on the node's user priority, which should be more than or equal to a randomly distributed number z from [0,1]. Different collision probabilities are allocated to devices based on their precedence (see Table 7.4). The CP is initially set to maximum for new nodes that have not previously received an allocation [24]. Thereafter the node follows the following algorithm for updating the CP value:

$$CP = \begin{cases} CP_{max} & \text{when no failure} \\ CP & \text{when odd number of failures} \\ 0.5 \times CP \text{ or } CP_{min} & \text{when even number of failures, depending on} \\ & \text{whichever is higher} \end{cases}$$

The flowchart for CP updating using this protocol is shown in Figure 7.5.

As can be seen in Figure 7.5, the Slotted Aloha protocol has the drawback that when contention probability is halved the chance to transmit a packet reduces, and this causes further delay in packet transmission. Modifications to Slotted Aloha are being developed by researchers to overcome these limitations.

Table 7.4 Bounded value for CSMA/CA and Slotted Aloha protocol

User Priority	Traffic Designation	Frame Type	Slotted Aloha		CSMA/CA	
			CP_{max}	CP_{min}	CW_{min}	CW_{max}
0	Background (BK)	D	0.125	0.0625	16	64
1	Best effort (BE)	D	0.125	0.0937	16	32
2	Excellent effort (EE)	D	0.25	0.0937	8	32
3	Video (VI)	D	0.25	0.125	8	16
4	Voice (VO)	D	0.375	0.125	4	16
5	Medical data or network control	D/M	0.375	0.1875	4	8
6	Urgent medical information	D/M	0.5	0.1875	2	8
7	Emergency information	D	1	0.25	1	4

D: Data, M: Management

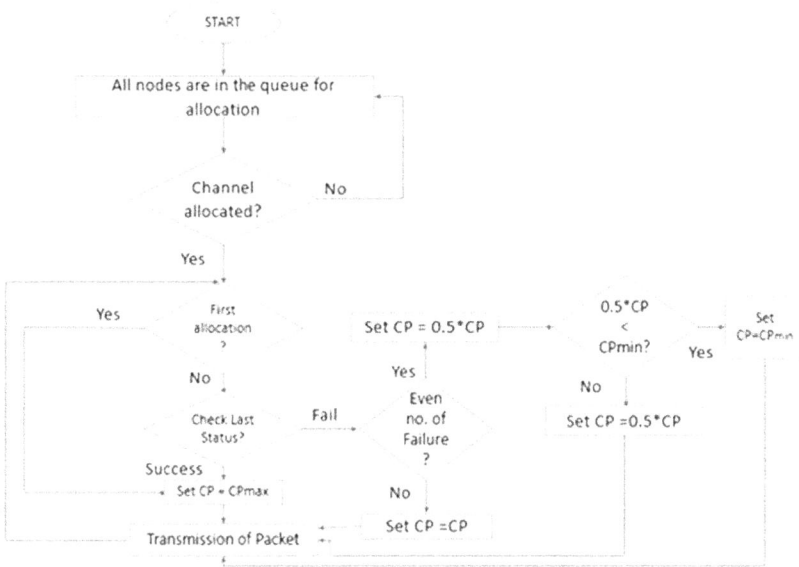

Figure 7.5 Flow chart of collision probability setting for Slotted Aloha.

A variety of traffic services, including high-effort and time-sensitive communication services, are expected to be made possible using various node priorities.

7.5.2 CSMA/CA access

This protocol deals with the two parameters for the data transmission-contention window (CW), and backoff counter (BC). With the help of this protocol, nodes can put their BCs to random values evenly distributed throughout the range [1, CW], where CW stands for the contention window and is chosen from the range [CW_{min}, CW_{max}]. The parameters CW_{min} and CW_{max} represent the lowest and highest contention window, respectively, and have different values for different user priorities (Table 7.4). The number of aborted attempts during information transmission decides the value of CW.

To start communication, clear channel assessment (CCA) is performed by the nodes to examine channel availability. CW is set to CW_{min} and stays there for every successful transfer. The BC is activated when the node detects the network during an unoccupied short interframe space (pSIFS) interval or when the current moment in the EAP, CAP, and RAP is adequate to handle the whole communication, as illustrated in Figure 7.6. When other data communication services use the channel and the permitted slot length is insufficient for data packet transfer, the BC is paused. When the failure-contention window for a node is altered, it doubles for even numbers of failures and stays the same for odd numbers of failures [14–21].

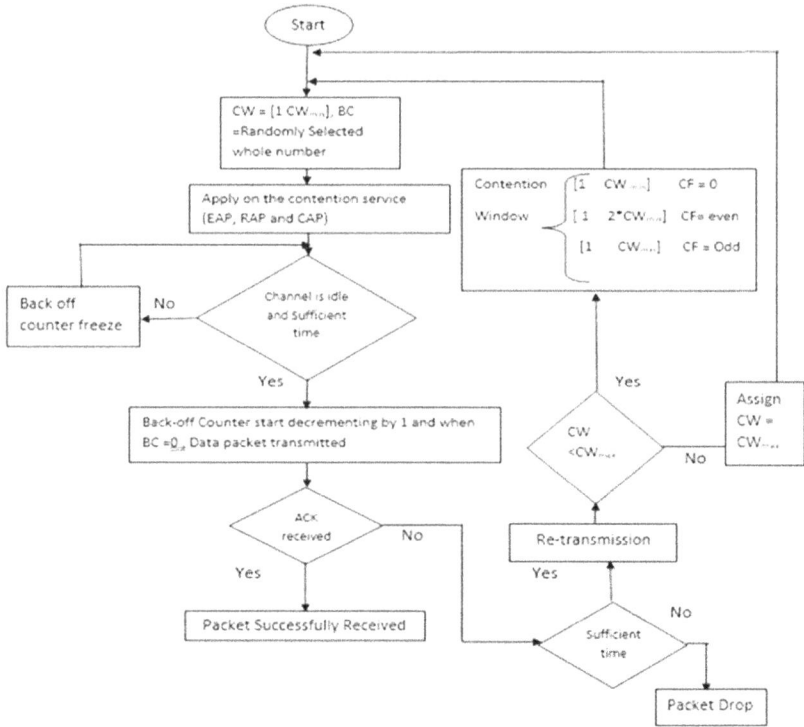

Figure 7.6 Flow chart of CSMA/CA process.

When a user node has a channel grant, the station can transmit up to two frames if the user priority is 0 to 5, or up to four frames if the user priority is 6 or more than 6.

7.6 RESEARCH AND DEVELOPMENT FOR MBAN

IEEE 802.15.6 allows different types of sensors to be connected at the node, and this makes the traffic heterogeneous. Traffic management in this situation becomes challenging. For the proper transmission of data to the target location, timely channel allocation to sensor nodes is essential; however, the fixed allocation of reserved slots to sensor nodes has some limitations and is irrelevant to actual or dynamic circumstances. As a result, different approaches for providing dynamic slot distribution have been developed, as shown in Table 7.5 [25–29]. To highlight the significance of QoS and resource utilization, Muthulakshmi and Shyamala (2017) suggest a MAC protocol that manages essential traffic using short slots in the beacon-enabled super-frame and regular traffic using a planned slots allocation strategy depending on data rate. The high energy usage of the nodes with high data rates is a major drawback of this technique.

Table 7.5 Summary of dynamic slot allocation techniques

Author name & year	Technique	Drawback
Muthulakshmi et al., 2017	Dynamic slot scheduling based on temporal autocorrelation	Node including a greater data rate consumes a significant quantity of energy
Salama et al., 2018	Slot scheduling based on the time division multiple access mechanism	Node prioritization is not considered while allocating slots to nodes
Zhang et al., 2018	Non-overlapping CW-based dynamic slots	Every node has the same transmit power
Saboor et al., 2020	DSA scheme using non-overlapping congestion window	Various healthcare situations were not addressed
Soni et al., 2021	VarySchedSlots scheme	Works with scheduled slots and does not evaluate various health issues

Additionally, Salayma et al. (2018) developed two-time division multiple access-based scheduling techniques to increase reliability and energy efficiency: adaptive scheduling and dynamic scheduling. The first technique dynamically assigns time slots to nodes by evaluating their network and storage health in both emergency and non-emergency situations. The second grants time slots to nodes based on their queue length. However, node prioritizing is not considered in these slot allocation algorithms.

Zhang et al. (2018) proposed a dynamic slot scheduling (DSS) technique for optimizing slot scheduling in the super-frame using a temporal autocorrelation model. Actual on-body data is examined to ascertain the unpredictability of communication networks obtained from certain wireless transceivers. The disadvantage of this system is that each node has the same signal strength.

Saboor et al. (2020) suggested a non-overlapping contention window-based dynamic slots allocation approach. This technique introduces two algorithms: the non-overlapping backoff algorithm (NOBA) and the dynamic slot allocation (DSA) scheme. The NOBA removes backoff-induced inter-priority conflicts, whereas the DSA distributes dynamic slots to decrease wastage caused by the fixed slot size of the super-frame structure. However, this system has not been evaluated in several healthcare scenarios. Finally, Soni et al. (2021) suggested the "VarySchedSlots" approach, which works with scheduled slots only with certain health conditions and does not address unscheduled traffic.

Table 7.6 summarizes some of the research done on improving the performance parameters for communication using this standard.

Table 7.5 summarizes the state of the art for dynamic access algorithms in the IEEE 802.15.6 standard, including assumptions made for the research. Many researchers are working on optimizing performance parameters and minimizing delays to achieve real-time data communication using MBAN.

Table 7.6 Summary of existing approaches [19, 21, 23, 30–33]

	Author and year	Access protocol	Assumption	Approach	Access phase	Traffic
Relay Mechanism	Ismail et al., 2016	CSMA /CA	Without contention	Algorithm for relay selection and simulation analysis	EAP 1, RAP1, MAP1	UP6 and UP7
	Michaelides et al., 2019	CSMA /CA	Rope Skipping scenario	Mobility aware relaying scheme	EP	UP7
	Zhang et al., 2020	MAC layer	Uplink transmission. In comparison to sensor nodes, BNH has sufficient energy, the movement sensor node obtains its movement-by-movement recognition method, and no energy is wasted on idle listening, overhearing, and collision	Adaptively relay allocation	MAP	Randomly
	Waheed et al., 2021	MAC layer	Stable contention channel	T-Relay protocol	RAP	On-demand, Emergency, Normal
Dynamic slot allocation and Prioritization	Bukvic et al., 2015	CSMA/CA	Unsecured communication	Simulation study	MAP	UP7
	Fourati et al., 2017	CSMA/CA	Saturated network	Algorithm for dynamic backoff bounds	EAP 1, RAP 1	All UP
	Das et al, 2021	CSMA/CA	Contention based	A mathematical model for BC value	EAP, RAP	Randomly

7.7 CONCLUSIONS

The emerging technology MBAN is commonly known as wearable or implantable healthcare. IEEE Standard 802.15.6 used for MBAN allows eight levels of user priorities for nodes, such that data from a high-priority node is transmitted with minimum delay. The MAC layer mechanism CSMA/CA and Slotted Aloha are designed so that the higher-priority users are granted a higher probability for channel access. Researchers are proposing newer mechanisms that result in less contention and higher throughput for heterogeneous traffic. Protocols for dynamic channel allocation, energy optimization, and relay mechanism are evolving, and the technology is expected to emerge as practical remote healthcare monitoring.

REFERENCES

[1] W.H. Lin, H. Wang, O. W. Samuel, G. Liu, Z. Huang, and G. Li, "New Photoplethysmogram Indicators for Improving Cuffless and Continuous Blood Pressure Estimation Accuracy," *Physiological Measurement*, vol. 39, no. 2, p. 025005, Feb. 2018, doi: 10.1088/1361-6579/aaa454.

[2] S. Pirbhulal, H. Zhang, M. E. E. Alahi, H. Ghayvat, S. C. Mukhopadhyay, Y. T. Zhang, and W. Wu, "A Novel Secure IoT-Based Smart Home Automation System Using a Wireless Sensor Network," *Sensors*, vol. 17, no. 3, p. 69, Dec. 2016, doi: 10.3390/s17010069.

[3] Z. Ali, M. Imran, M. Alsulaiman, M. Shoaib, and S. Ullah, "Chaos-Based Robust Method of Zero-Watermarking for Medical Signals," *Future Generation Computer Systems*, vol. 88, pp. 400–412, Nov. 2018, doi: 10.1016/j.future.2018. 05.058.

[4] D. Lin, Y. Tang, F. Labeau, Y. Yao, M. Imran, and A. V. Vasilakos, "Internet of Vehicles for E-Health Applications: A Potential Game for Optimal Network Capacity," *IEEE System Journal*, vol. 11, no. 3, pp. 1888–1896, Sep. 2017, doi: 10.1109/JSYST.2015.2441720.

[5] A. Kamble and S. Bhutad, "IOT Based Patient Health Monitoring System with Nested Cloud Security," in *2018 4th International Conference on Computing Communication and Automation (ICCCA)*, Greater Noida, India, Dec. 2018, pp. 1–5, doi: 10.1109/CCAA.2018.8777691.

[6] B. Pradhan, S. Bhattacharyya, and K. Pal, "IoT-Based Applications in Healthcare Devices," *Journal of Healthcare. Engineering.*, vol. 2021, pp. 1–18, Mar. 2021, doi: 10.1155/2021/6632599.

[7] R. M. Gardner, T. P. Clemmer, R. S. Evans, and R. G. Mark, "Patient-Monitoring Systems," in *Biomedical Informatics*, E. H. Shortliffe, and J. J. Cimino, (eds) Springer, London. 2014, pp. 561–591. doi: 10.1007/978-1-4471-4474-8_19.

[8] M. L. Ranney, V. Griffeth, and A. K. Jha, "Critical Supply Shortages — The Need for Ventilators and Personal Protective Equipment during the Covid-19 Pandemic," *The.New England Journal of Medicine*, vol. 382, no. 18, p. e41, Apr. 2020, doi: 10.1056/NEJMp2006141.

[9] K. Dadhich, V. Dadhich, and D. Kataria, "Healthcare Monitoring Using Wireless Sensors," in *2022 IEEE Delhi Section Conference (DELCON)*, New Delhi, India, 2022, pp. 1–6, doi: 10.1109/DELCON54057.2022.9753456.

[10] J. E. Hollander and B. G. Carr, "Virtually Perfect? Telemedicine for Covid-19," *The New England Journal of Medicine*, vol. 382, no. 18, pp. 1679–1681, Apr. 2020, doi: 10.1056/NEJMp2003539.

[11] M. Sabov and C. E. Daniels, "The Value of 24/7 In-House ICU Staffing 24/7 Intensivist in the ICU," *Critical Care Medicine.*, vol. 46, no. 1, pp. 149–151, Jan. 2018, doi: 10.1097/CCM.0000000000002747.

[12] H. S. Ahmed and A. A. Ali, "Smart Intensive Care Unit Desgin Based On Wireless Sensor Network and Internet of Things," in *2016 Al-Sadeq International Conference on Multidisciplinary in IT and Communication Science and Applications (AIC-MITCSA)*, Baghdad, Iraq, May 2016, pp. 1–6. doi: 10.1109/AIC-MITCSA.2016.7759905.

[13] R. A. Khan and A.-S. K. Pathan, "The State-Of-The-Art Wireless Body Area Sensor Networks: A Survey," *International Journal of Distributed Sensor Networks*, vol. 14, no. 4, pp. 1–23, Apr. 2018, doi: 10.1177/1550147718768994.

[14] T. Benmansour, T. Ahmed, S. Moussaoui, and Z. Doukha, "Performance analyses of the IEEE 802.15.6 Wireless Body Area Network with heterogeneous traffic," *Journal of Network and Computer Application.*, vol. 163, p. 102651, Aug. 2020, doi: 10.1016/j.jnca.2020.102651.

[15] X. Yuan, C. Li, Q. Ye, K. Zhang, N. Cheng, N. Zhang, and X. Shen, "Performance Analysis of IEEE 802.15.6-Based Coexisting Mobile WBANs With Prioritized Traffic and Dynamic Interference," *IEEE Transaction on Wireless Communication.*, vol. 17, no. 8, pp. 5637–5652, Aug. 2018, doi: 10.1109/TWC.2018.2848223.

[16] S. Ullah, M. Mohaisen, and M. A. Alnuem, "A Review of IEEE 802.15.6 MAC, PHY, and Security Specifications," *International Journal of Distributed Sensor Network*, vol. 9, no. 4, p. 950704, Apr. 2013, doi: 10.1155/2013/950704.

[17] M. Hernandez and L. Mucchi, "Survey and Coexistence Study of IEEE 802.15.6™ -2012 Body Area Networks, UWB PHY," in *Body Area Networks Using IEEE 802.15.6*, M. Hernandez and L. Mucchi (eds) Academic Press 2014, pp. 1–44. doi: 10.1016/B978-0-12-396520-2.00001-7.

[18] R. Huang, Z. Nie, C. Duan, Y. Liu, L. Jia, and L. Wang, "Analysis and Comparison of the IEEE 802.15.4 and 802.15.6 Wireless Standards Based on MAC Layer," in *Health Information Science*, Springer International Publishing vol. 9085, 2015, pp. 7–16. doi: 10.1007/978-3-319-19156-0_2.

[19] T. Waheed, A. U. Rehman, and F. K. Shaikh, "IEEE 802.15.6 Relaying Protocol for MBANs," in *2021 Mohammad Ali Jinnah University International Conference on Computing (MAJICC)*, Karachi, Pakistan, Jul. 2021, pp. 1–6. doi: 10.1109/MAJICC53071.2021.9526263.

[20] A. Vyas, S. Pal, and B. K. Saha, "Relay-Based Communications in WBANs: A Comprehensive Survey," *ACM Computing Surveys (CSUR)*, vol. 54, no. 1, pp. 1–34, Jan. 2021, doi: 10.1145/3423164.

[21] Y. Zhang, B. Zhang, and S. Zhang, "An Adaptive Energy-Aware Relay Mechanism for IEEE 802.15.6 Wireless Body Area Networks," *Wireless Personal Communication*, vol. 115, pp. 2363–2389, Dec. 2020, doi: 10.1007/s11277-020-07686-4.

[22] T. Benmansour, T. Ahmed, and S. Moussaoui, "Performance Analyses and Improvement of the IEEE 802.15.6 CSMA/CA using the Low Latency Queuing," in *2017 IEEE 22nd International Workshop on Computer Aided Modeling and Design of Communication Links and Networks (CAMAD)*, Lund, Sweden, Jun. 2017, pp. 1–6. doi: 10.1109/CAMAD.2017.8031623.

[23] H. Fourati, H. Idoudi, and L. A. Saidane, "Intelligent Slots Allocation for Dynamic Differentiation in IEEE 802.15.6 CSMA/CA," *Ad Hoc Networks*, vol. 72, pp. 27–43, Apr. 2018, doi: 10.1016/j.adhoc.2018.01.007.

[24] M. Fatehy and R. Kohno, "A Novel Contention Probability Dynamism for IEEE 802.15.6 Standard," *EURASIP Journal on Wireless Communications and Networking*, vol. 2014, no. 1, pp. 1–10, Dec. 2014, doi: 10.1186/1687-1499-2014-92.

[25] A. Muthulakshmi and K. Shyamala, "Efficient Patient Care Through Wireless Body Area Networks—Enhanced Technique for Handling Emergency Situations with Better Quality of Service," *Wireless Personal Communications*, vol. 95, no. 4, pp. 3755–3769, Aug. 2017, doi: 10.1007/s11277-017-4024-7.

[26] M. Salayma, A. Al-Dubai, I. Romdhani, and Y. Nasser, "Reliability and Energy Efficiency Enhancement for Emergency-Aware Wireless Body Area Networks (WBANs)," *IEEE Transactions on Green Communications and Networking*, vol. 2, no. 3, pp. 804–816, Mar 2018, doi: 10.1109/TGCN.2018.2813060.

[27] H. Zhang, F. Safaei, and L. C. Tran, "Channel Autocorrelation-Based Dynamic Slot Scheduling for Body Area Networks," *EURASIP Journal on Wireless Communications and Networking.*, vol. 2018, pp. 1–17, Dec. 2018, doi: 10.1186/s13638-018-1261-8.

[28] A. Saboor, R. Ahmed, W. Ahmed, A. K. Kiani, M. M. Alam, A. Kuusik, and Y. Le Moullec, "Dynamic Slot Allocation Using Non Overlapping Backoff Algorithm in IEEE 802.15.6 WBAN," *IEEE Sensors Journal*, vol. 20, no. 18, pp. 10862–10875, Sep. 2020, doi: 10.1109/JSEN.2020.2993795.

[29] G. Soni and K. Selvaradjou, "A Dynamic Allocation Scheme of Scheduled Slots for Real-Time Heterogenous Traffic in IEEE 802.15.6 Standard for Scheduled Access Mechanism," *Journal of Ambient Intelligence and Humanized Computing*, vol. 14, no. 1, pp. 237–256, May 2021, doi: 10.1007/s12652-021-03288-5.

[30] M. Ismail, F. Bashir, Y. Zia, S. Kanwal, and M. E. Azhar "Relaying Node Selection Technique For IEEE 802.15.6," *Journal of Information Security Research* vol. 7, no. 3, pp. 91-100, Sept. 2016.

[31] C. Michaelides, M. Iloridou, and F.-N. Pavlidou, "An Improved Mobility Aware Relaying Scheme for Body Area Networks," *IEEE Sensors Journal*, vol. 19, no. 16, pp. 7141–7148, Aug. 2019, doi: 10.1109/JSEN.2019.2912892.

[32] M. Bukvic and J. Misic, "Traffic Prioritisation in 802.15.6 MAC: Analysis of Its CSMA/CA Algorithm and Proposals for Improvement," *Telfor Journal*, vol. 7, no. 1, pp. 8–13, July 2015, doi: 10.5937/telfor1501008B.

[33] K. Das and S. Moulik, "PBCR: Parameter-based Backoff Counter Regulation in IEEE 802.15.6 CSMA/CA," in *2021 International Conference on Communication Systems & NETworkS (COMSNETS)*, Bangalore, India, Jan. 2021, pp. 565–571. doi: 10.1109/COMSNETS51098.2021.9352747.

Chapter 8

Simultaneous encryption and compression for securing large data transmission over a heterogeneous network

Shiladitya Bhattacharjee
UPES, Dehradun, India

Sulabh Bansal
Manipal University Jaipur, Jaipur, India

CONTENTS

8.1 INTRODUCTION

Data storage in a physical medium or transmitted via a delivery channel among the computers within heterogeneous communication networks involves repetition. Data compression minimizes prolixity to extricate physical disk space and reduce transportation time [1]. Different pattern libraries

DOI: 10.1201/9781003415466-8

require various compression techniques that act specifically when the sender and receiver pair realize the information belongs to the library concerned. Since compression does not require a secret key or identification constraint during confining or deconfining [1, 2], the compressed data can be vulnerable to illegitimate access. The alternative is to use data encryption to convert and keep the data unreadable and unaltered to achieve data security. The random bits are generated from the input data blocks in the encryption mechanism using cryptic functions and secret keys.

Encryption can be performed before compression or vice versa. The main idea of compressing data is to reduce data repetition and encryption, improving data security against various statistical attacks [3, 4]. If encryption is performed first it produces cogent randomness with little data redundancy, rendering subsequent compression ineffective. Performing encryption and compression consecutively minimizes the storage and bandwidth requirements, and improves data security, but it is a slow and computationally expensive exercise. Therefore, this research work primarily focuses on recent results in chaos-based cryptography and compression and on developing an efficient integrated technique to address these security shortcomings and the issues related to data size.

8.1.1 Research scope

The present research work looks at developing parallel data compression and encryption algorithms. Here, the compression of data is conducted along with a secret key, enabling data encryption and compression to be executed in a single step. The proposed technique can be applied to any data type and size. The concurrent mechanism incorporates randomness into the compressed data and enhances encryption quality. Compression and encryption are executed in such a way that the outcome can only be decoded using the same secret key.

8.1.2 Research objectives

The specific objectives of this research are:

1. To develop a tightly integrated compression and encryption technique to control security issues with previous methods.
2. To reduce the time and space complexities.
3. To analyze the performance of the proposed technique against similar works.

8.1.3 Organization of the paper

The remaining sections are as follows. Section 8.2 discusses the strengths and limitations of the existing security techniques that apply to extensive data transmission in the light of recent security problems discovered in large data transmissions. Section 8.3 describes the simultaneous encryption and compression technique and how compressed and encrypted files are decoded. The experimental setup and data preparation is discussed in Section 8.4, together with the assessment parameters used to evaluate the technique. Section 8.5 compares this technique with existing methods and demonstrates its superior achievements. Finally, Section 8.6 considers the effectiveness and shortcomings of this research, and possible future research directions.

8.2 LITERATURE REVIEW

The many threats to busy networks – from insufficient storage capacity, incursions and network susceptibility to illicit interference, diverse security threats and the problems of big data [5] – have led to a surge in interest in concurrent compression and encryption applications. The integrated application of compression and encryption targets offers data security while reducing data repetition. To date, research has sought to combine encryption competence with regular data compression applications such as Huffman compression and arithmetic compression [6, 7].

Research in [7, 8] proposes a combined compression and encryption technique that replaces the branches of the Huffman Tree (HT) from left to right or vice versa with the relevant control key. All the adapted trees created with this technique were rigid in size. A unique and improved collective Huffman Tree-based approach [6, 7, 9] was advocated to manage the fixed-size problem, along with a statistical model-oriented compression for creating adapted tables. A proposed improvement used concurrent reformed HTs in [9] that overcame the various codeword problems to augment the key space and determine the weak security contentions. This approach handles the tree enhancements by applying a pseudorandom keystream created by a simultaneous map. A proposed updated form of traditional arithmetic compression used a regulatory key to control the simultaneous compression and encryption, but this was susceptible to common-plaintext incursion [9].

In [8, 10], an arbitrary arithmetic coding approach for the JPEG 2000 format was proposed, comprising randomization protocols for the traditional arithmetic compression applications for encryption. Interval swapping achieves compression and encryption during the encoding and decoding process, which is controlled by generating a random sequence using the control key. Furthermore, the conventional adaptive arithmetic coding was modified with a secret key mechanism based on chaotic PRKG to control the statistical model used in the compression process [8].

Many combined compression and encryption algorithms have been designed by introducing cryptographic capabilities into conventional compression techniques such as arithmetic coding and Huffman coding [11]. However, these algorithms still pose serious security issues, and their vulnerability is exposed in [12], which indicates that the combined effect of chaos, cryptography, and compression had not been comprehensively examined. The chaotic structures were applied when pseudorandom bitstream creators comprised the primary reins.

Concurrent data compression arrangements were proposed in [13, 14], which incorporates the Bernoulli shift map as a piecewise linear Markov map (PWLM) in the probability distribution of source directives. It finds the optimal primary estate for repeating a chaotic map to produce a symbolic flow that expresses the source message. A nonlinear dynamic approach for chaotic systems, the generalized Luroth series (GLS), was introduced and ratified to be optimal according to the Shannon theorem [11–13]. However, the complexity of these techniques renders them unfeasible for extensive input data.

8.2.1 Research gap analysis

Compression and encryption of data are independent restraints with their own advantages and disadvantages; their construction is intricate and requires comprehensive execution capacity to control big datasets. This section discusses their essential functionalities and performance from the compression and security perspective. When both compression and security are required, these two processes are employed sequentially on the same datasets to reduce data size and protect data from unethical use or access during storage and transmission. In practice this is an either/or process: the data is either compressed prior to encryption or encrypted prior to compression [15, 16]. The performance of both these approaches has been evaluated in terms of ability to compress while maintaining data security. Background studies have found that compression-first strategies adequately reduce the required storage space and transportation overheads, as well as improving data security. However, this approach requires the result of one application to be fed to the other, a convoluted and time-consuming process.

According to background studies, these existing techniques offer the same compression efficiency as conventional algorithms. However, the literature review exposes their many limitations, extremes of performance, and security vulnerabilities. Moreover, the complexities of the combination of chaos, encryption, and compression has yet to be thoroughly examined. The chaos approach was exclusively applied as a pseudorandom bitstream generator incorporating critical controls in HT mutation or AC interval splitting [17]. Although many simultaneous data compression and encryption techniques have been applied to address these complexities, the fundamental issues remain to be solved [16]. No suitable method is yet available that can address the compression and encryption issues in an integrated way without one affecting the other or creating a new problem. Hence the requirement for new techniques that meet all the data storage and communication system criteria without compromising any fundamental aspect of compression and security.

8.3 PROPOSED TECHNIQUE

The proposed technique comprises three primary parts: chaotic S-Box, secure and robust Huffman coding, and a pseudorandom keystream generator (PRKG). The projected approach incorporated CLM (chaotic logistic map) for essential restraint into compression and decompression techniques of secure adaptive Huffman coding (AHC). The chaotic S-Box is based on a chaotic sine map (CSM) for conducting data replacement without adjusting data compression abilities. Figure 8.1 is a basic flow diagram of the projected concurrent secure data compression (CSDC) system, including the arrangements for achieving data compression and decompression.

Inputs and outcomes: The concurrent encryption and data compression are conducted by incorporating the trilateral confidential keys $K1$, $K2$, and $K3$ (every secret key is the consolidation of the opening appraisal $x0$ as well as regulation ambit λ to their corresponding chaotic map). The use of $K1$ is included in CSM, which builds an aggressive S-Box, $K2$, as well as $K3$. These components are applied to create the pseudorandom keystreams with the help of CLM. In this deck, the input data I, length L, and the resultant C (compressed and encrypted data) are used in the following Algorithm 1.

ALGORITHM I: Concurrent Compression and Encryption

1. **Set up** the primary Adaptive Huffman Tree (AHT).
2. **Develop** a chaotic S-Box with the help of the confidential key $K1$.
3. **Trigger** the counter's value as $i = 1$ and the resultant data D as blank.
4. **Scan** the symbol I_i as input from I, represents as input data.
5. **Alternate** I_i as the input symbol with the help of a chaotic S-Box where s_i represents the resultant symbol.
6. **Convert** s_i with the help of encryption, robust Huffman coding technique, and critical $K2$. Integrate the resultant sting ci along with the D as output data as $D = D \parallel c_i$ (here, token "$\parallel$" enacts the integration activity).
7. **Raise** the value of the t counter as $i = i + 1$.
8. The secret key $K3$ hatches the pseudorandom keystream KS with the help of PRKG, depending upon the CLM. Furthermore, it masks the entire resultant data D. The application of masking is accomplished along with an XOR operation, which amplifies the complete randomness of resultant data D. This part generates the decisive encrypted text C such as $C = D \otimes KS$.
9. **Test** when $i > L$; repeat Step 8. Otherwise, repeat Step 4.
10. The construction of secure data compression ends here.

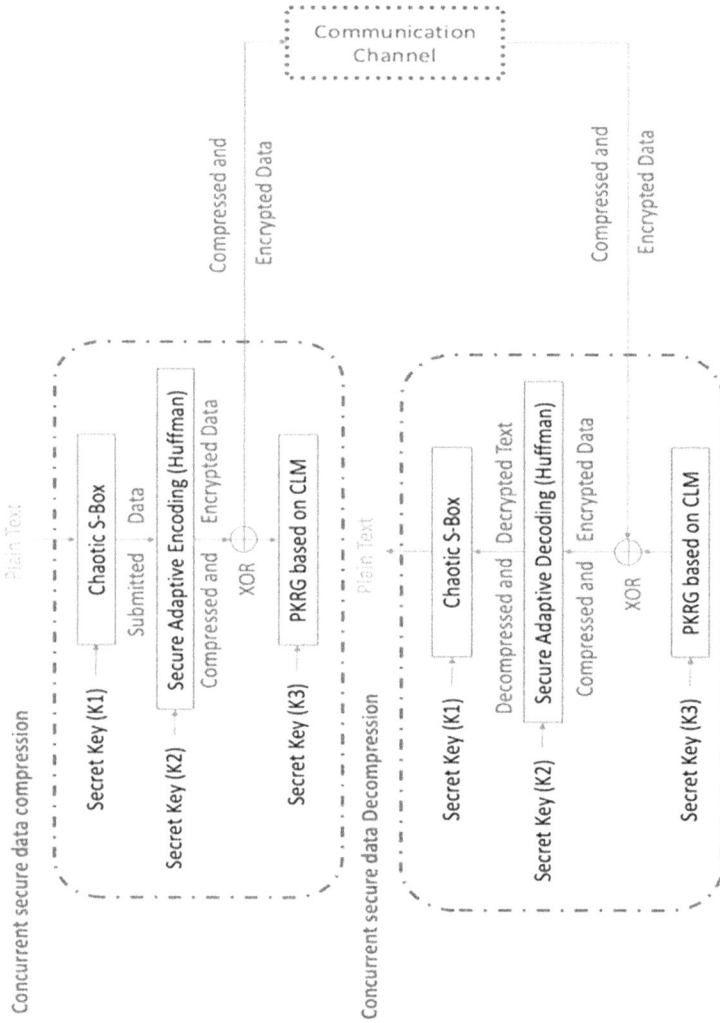

Figure 8.1 Fundamental flow structure of the proposed technique.

The decompression, along with the security, is like the cover of a particular compression technique. It needs to duplicate the triode secret keys $K1$, $K2$, and $K3$ for the encoding. Primarily, it performs decryption operations with the cryptic text with the help of PRKG, depending upon the CLM and the confidential key $K3$. It untangles the data with the help of a secure and robust Huffman decoder with the confidential key $K2$. Conduct the reverse substitution adopting the chaotic S-Box together with the confidential key $K1$ after that. It offers the actual plaintext. This development of decompression in a secure manner is shown in below Algorithm 2.

Inputs and outcomes: The decompression of data in a secure way is conducted by replicating the triode secret keys $K1$, $K2$, and $K3$ (every secret key is the consolidation of the opening appraisal $K0$ as well as regulation ambit λ to their corresponding chaotic map) similar to the application of the secret keys in the secure compression of data. Here the input cryptic text is represented by C, and the resultant plaintext is represented by P.

ALGORITHM 2: Concurrent Decompression and Decryption

1. Applying the PRKG, create a keystream (KS) depending upon CLM along with $K3$. Employ the new KS for unmasking the cryptic text C and to generate the compressed data D such as $D = C \otimes KS$.
2. Construct the primary AHT.
3. Create the chaotic S-Box with the help of confidential key $K1$.
4. Create the plaintext P and initialize it as empty.
5. Scan the c, the compressed code string, from D, the compressed data string.
6. Untangle the compressed code c using the protected and robust Huffman coding and together with key $K2$ within the resultant code s.
7. Replace the resultant string s in the outcome I with the help of a chaotic S-Box. Integrate I with P such as $P = P \parallel I$ (here, token " $\parallel$ " enacts the integration activity).
8. Stop the decompression process if the file ends. Otherwise, repeat Step 5.
9. The construction of fast decompression ends here.
   Thus, the protected data compression and decompression can be performed with the input plain text with Algorithms 1 and 2.

8.4 ASSESSMENT PLATFORM

This section describes the basis of the experiment to construct the projected combined technique and its various components. The download sources of different compulsory files as inputs, various software, and the creation of appropriate input data are also defined. Finally, this section explains the performance evaluation matrices applied to assessment of the proposed combined approach against existing security techniques. These matrices are categorized according to their effectiveness in the context of this research's specific objectives.

8.4.1 Experiment setup

This part consists of two sub-sections: system requirement, which describes data preparation and the hardware and software setup for executing the proposed technique; and the details of the data preparation process.

8.4.1.1 System requirement

The planned approach and its corresponding components were constructed with the help of Java (JDK 7.0). The 16.04 LTS version of the UBUNTU operating system was used to conduct various experiments. Parallel processing was accelerated with Java thread programming. The PC used was configured with DDR3 RAM of 32GB and i8 processors from Intel® Core™. All files were transferred during the experimental work via wireless and wired local area networks.

8.4.1.2 Data preparation

According to [18, 19], the standard version of Calgary Corpus is usually used as the yardstick to appraise any data compression mechanism for inputs, in order to maintain standardization. Therefore, the standard version of Calgary Corpus was applied as input files in this research. In addition, text and binary files of different sizes up to 1TB were generated during the experiment and were employed as inputs to test the performance of the proposed compression technique. As Java can process a maximum of 64MB of data at a time, a file splitter was designed to split the secret input file into several small 64MB-sized parts. The splitting operation was carried out to accelerate the overall process from the sending end. Similarly, a file merger was designed to merge the received small files into a single large file at the receiving end to generate a complete output file. During the splitting operation, each small file was assigned an index number. These index numbers were used to merge small files into a single large file at the receiving end.

8.4.2 Assessment parameters

The proposed integrated technique was designed to assess different aspects of the proposed concurrent compression and encryption performance. These vital parameters for measuring the effectiveness of the proposed integrated approach and its various parts are presented and defined according to their capacity to fulfill the stated objectives of this research in the following sub-section.

8.4.2.1 Randomness analysis using the NIST

In this research work, we formed binary sequences of 300 along with the $L = 1,000,000$ bits to measure the cryptic characteristics of pseudorandom keystreams created with the issues of PRKGs. The regulation ambit values λ were considered arbitrarily, where the values of λ lie between 3.6 and 4 for the CLM. Another initial value X0 was randomly considered between 0 and 1 for the generator pair. In favor of confirming the randomness certainty as 99% of pseudorandom sequences of binary strings formed with the help of PRKG depending upon CLM [19, 20], the level of importance (α) was accommodated to 0.01.

8.4.2.2 Throughput

In a computing system, any job is accomplished within a specified time frame. The time required to perform different jobs varies with the computing system's processing speed and the nature of the job. Similarly, time is essential in the data transmission system, as the actual data must be transmitted within a specified time. Data loss may occur if any unwanted time delay occurs during the transportation. Hence, the performance of any data transmission system can be measured by its time requirement. According to [14–16], *throughput* is the amount of a task done in a specific time. The throughput (TP) produced by any technique can be calculated as

$$TP = \left(\frac{\text{Output file size}}{\text{Total execution time to generate output}} \right) \qquad (8.1)$$

8.4.2.3 Percentage of space saving

According to the literature, compression techniques are applied to any data file to save disk storage space requirements or reduce the bandwidth requirement for transmission over the network. Space saving can be defined as the saving of space or bandwidth requirement of any compressed file for storage or transmission after any special compression operation has been performed with an input file [8, 10]. It is generally measured in percentage terms. The space-saving rate offered by any fixed-length or variable-length coding compression technique usually varies according to the input file sizes. The space-saving rate achieved by any compression technique can be calculated as

$$\text{Space saving}(\%) = \left(1 - \frac{\text{Compressed File Size}}{\text{Uncompressed File Size}} \right) \times 100 \qquad (8.2)$$

8.5 RESULT ANALYSIS

The background review shows that any data compression approach that offers a higher percentage of space saving is robust and effective in saving disk space and minimizing transportation overheads [15–17]. Similarly, the time efficiency and randomness (to ensure its capacity for providing data security) have also been tested in this section. Table 8.1 lists the percentages for the entire analyses with delinquency input ambits settings explicitly described in the NIST statistic test suite.

Both PRKGs generated pseudorandom keystreams that fortuitously cleared the entire NIST SP800-22 analysis, with the bounds ranging from 97.28% to 99.93%. Furthermore, the statistical assessment shows that the newly created entire keystreams were independently circulated with 99% confidence. Figure 8.2 shows the time efficiencies for compression with encryption for different file sizes based on Equation (8.1).

Figure 8.2 shows that the projected combined technique has more adequate time to produce higher throughputs than the other encryption and compression combinations. The proposed method's outputs are also compared with additional related decompression and decryption techniques using Equation (8.1) and plotted in Figure 8.3.

It can be seen from Figure 8.3 that the projected integrated approach offers higher time efficiency while offering higher throughputs than the corresponding existing decompression and decryption combinations. Furthermore, the space-saving efficiencies of the planned and distinct compression approaches have been calculated with the help of Equation (8.2) and are compared with the others in Figure 8.4.

Table 8.1 Percentages for all tests with default input parameters settings defined in NIST

Test	Passing % of pseudorandom keystreams generation (chaotic logistic map)
Frequency	99.67
Block frequency	98.67
Runs	100
The rank of binary matrix	99.33
Ones related to long runs	99.33
Templates for no overlapping	98.67
DFT spectral	97.33
Universal	98.33
Linear complexity	98.33
Serial	99.33
Approximate entropy	99
Cumulative sums	99.93
Random sum	99.33
Random excursions variant	98.67

Compression and/or Encryption Throughput

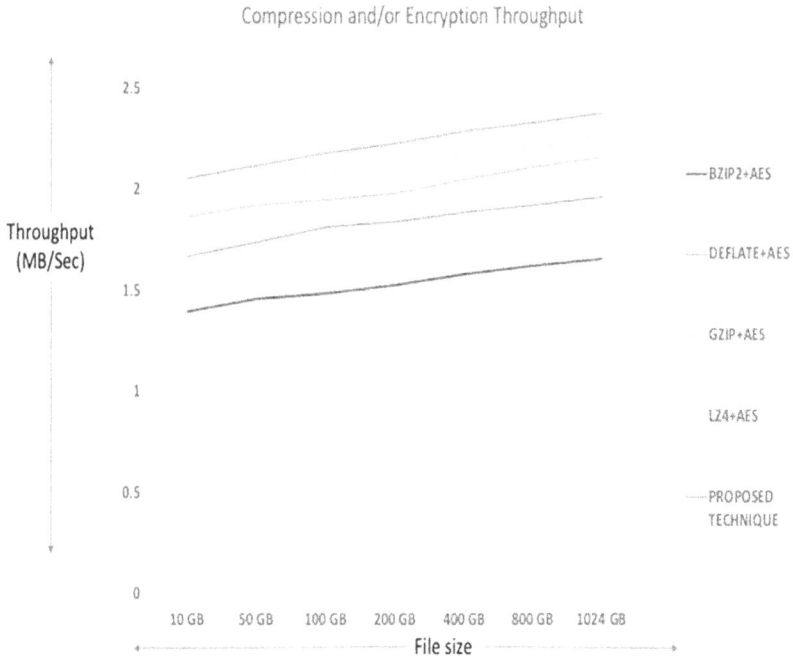

Figure 8.2 Time performance of different encryption and compression combinations.

Decompression and/or Decryption Throughput

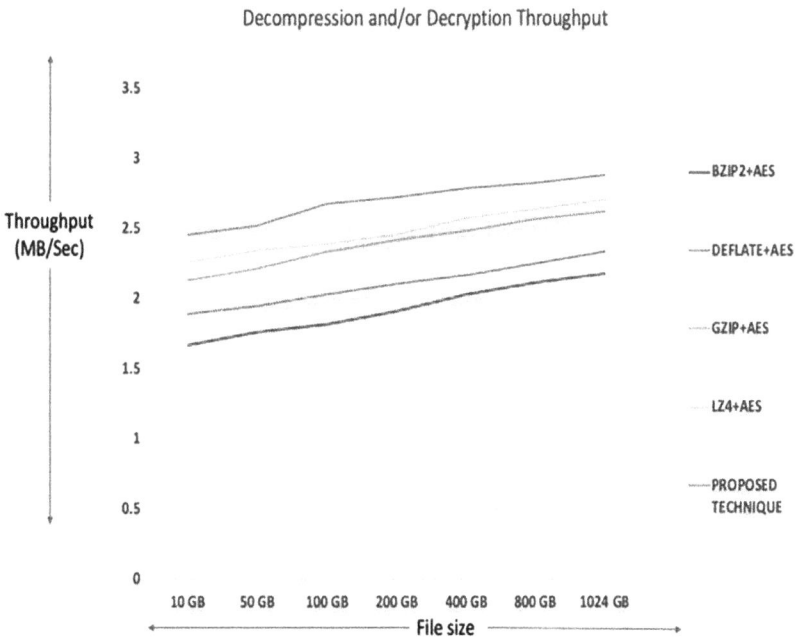

Figure 8.3 Time performance of different decryption and decompression combinations.

METOCEAN Data of Size 1 TB

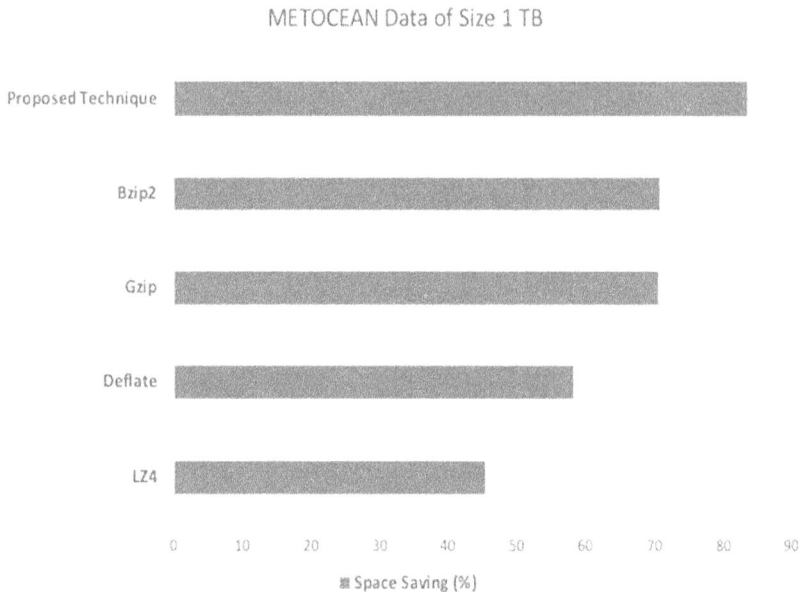

Figure 8.4 Space-saving efficiencies offered by different compression techniques.

From Figure 8.4, we can see that the proposed technique is reasonably efficient in terms of space-saving percentage with the help of Equation (8.2). The preliminary result shows that the planned combined approach is adequate in offering higher security as it offers higher randomness (shown in Table 8.1) as well as higher time and time efficiencies (Figures 8.2 and 8.3) and space-saving percentage (Figure 8.4).

8.6 CONCLUSION AND FUTURE WORK

The study confirms the advantages of the proposed technique compared to conventional and existing methods. It resolves the issues of time consumption for processing extensive data with the help of concurrent encryption and compression. The proposed technique also offers higher entropy values and space efficiencies than other correlated processes, efficiently reduces space and time requirements, and at the same time protects data confidentiality and integrity.

Future work will focus on improvements to the proposed techniques and other possible simultaneous data compression and encryption techniques based on chaos theory. Future research and development of the proposed work are not limited to compression efficiency but can also include cryptographic aspects. Chaos-based PRKGs were presented here. However, this area needs more investigation for better security and computational complexity since pseudorandom sequences are produced by combining two or more bits. Thus, the possibility of reducing the required number of bits without compromising safety is worth investigating.

REFERENCES

1. C. P. Wu and C. C. J. Kuo, "Design of integrated multimedia compression and encryption systems," *IEEE Trans. Multimed.*, vol. 7, no. 5, pp. 828–839, 2005.
2. S. Bhattacharjee, D. Midhun Chakkaravarthy, M. Chakkaravarthy, and L. B. A. Rahim "An integrated technique to ensure confidentiality and integrity in data transmission through the strongest and authentic hotspot selection mechanism." In *Data Management, Analytics and Innovation*, Springer, Singapore, vol-1016, pp. 459–474, 2020.
3. H. Hermassi, R. Rhouma, and S. Belghith, "Joint compression and encryption using chaotically mutated Huffman trees," *Commun. Nonlinear Sci. Numer. Simul.*, vol. 15, no. 10, pp. 2987–2999, 2010.
4. S. A. El-Said, K. F. A. Hussein, and M. M. Fouad, "Securing multimedia transmission using optimized multiple Huffman tables technique," *Int. J. Signal Process. Image Process. Pattern Recognit.*, vol. 4, no. 1, pp. 49–64, 2011.
5. S. Bhattacharjee, L. B. A. Rahim, A. W. Ramadhani, and D. Midhunchakkravarthy. "A study on seismic big data handling at seismic exploration industry", *Intelligent Computing and Innovation on Data Science*, Springer, Singapore, vol-118 pp. 421–429, 2020.
6. J. Zhou and O. C. Au, "Comments on a novel compression and encryption scheme using variable model arithmetic coding and coupled chaotic system," *IEEE Trans. Circuits Syst. I Regul. Pap.*, vol. 55, no. 10, pp. 3368–3369, Nov. 2008.
7. B. Mi, X. Liao, and Y. Chen, "A novel chaotic encryption scheme based on arithmetic coding," *Chaos Solit. Fractals*, vol. 38, no. 5, pp. 1523–1531, 2008.
8. S. Bhattacharjee, L. B. A. Rahim, and I. B. A. Aziz, "A lossless compression technique to increase robustness in big data transmission system", *Int. J. Adv. Soft Comput. Appl.*, vol. 7, no. 3, pp. 126–145, 2015.
9. J. G. Wen, H. Kim, and J. D. Villasenor, "Binary arithmetic coding with key-based interval splitting," *IEEE Signal Process. Lett.*, vol. 13, no. 2, pp. 69–72, 2006.
10. G. Jakimoski and K. P. Subbalakshmi, "Cryptanalysis of some multimedia encryption schemes," *IEEE Trans. Multimed.*, vol. 10, no. 3, pp. 330–338, 2008.
11. S. Bhattacharjee, L. B. A. Rahim, and I. B. Aziz, "Enhancement of confidentiality and integrity during big data transmission using a hybrid technique", *ARPN J. Eng. Appl. Sci.*, vol. 10, no. 23, pp. 18029–18038, 2015.
12. J. Zhou, O. C. Au, and P. H. W. Wong, "Adaptive chosen-ciphertext attack on secure arithmetic coding," *IEEE Trans. Signal Process.*, vol. 57, no. 5, pp. 1825–1838, 2009.
13. N. Nagaraj, P. G. Vaidya, and K. G. Bhat, "Arithmetic coding as a non-linear dynamical system," *Commun. Nonlinear Sci. Numer. Simul.*, vol. 14, no. 4, pp. 1013–1020, 2009.
14. S. Bhattacharjee, L. B. A. Rahim, and I. B. Aziz, "A security scheme to minimize information loss during big data transmission over the internet". *IEEE, 3rd International Conference on Computer and Information Sciences (ICCOINS)*, Kuala Lumpur, Malaysia, pp. 215–220, 2016.

15. K.-W. Wong, Q. Lin, and J. Chen, "Simultaneous arithmetic coding and encryption using chaotic maps," *IEEE Trans. Circuits Syst. II Express Briefs*, vol. 57, no. 2, pp. 146–150, 2010.
16. M. B. Luca, A. Serbanescu, and G. Burel, "A new compression method using a chaotic symbolic approach," *IEEE-Communications*, Bucharest, Romania, no. 1, pp. 1–6, 2004.
17. S. Puangpronpitag, P. Kasabai, and D. Pansa, "An enhancement of the SDP Security Description (SDES) for key protection," *IEEE, 9th International Conference on Electrical Engineering/Electronics, Computer, Telecommunications and Information Technology*, Phetchaburi, Thailand, pp. 1–4, 2012.
18. S. Bhattacharjee, L. B. A. Rahim, and I. B. Aziz, "A secure transmission scheme for textual data with least overhead", *IEEE, Twentieth National Conference on Communications (NCC)*, Kanpur, India, pp. 1–6, 2014.
19. R. Starosolski, "New simple and efficient color space transformations for lossless image compression," *J. Vis. Commun. Image Represent.*, vol. 25, no. 5, pp. 1056–1063, Jul. 2014.
20. S. K. Chen, "A module-based LSB substitution method with lossless secret data compression," *Comput. Stand. Interfaces*, vol. 33, no. 4, pp. 367–371, Jun. 2

Chapter 9

2D network on chip

Shreyash Yadav and Abhishek Sharma
The LNM Institute of Information Technology, Jaipur, India

Sulabh Bansal
Manipal University Jaipur, Jaipur, India

CONTENTS

9.1 INTRODUCTION

In recent years, the internet of things (IoT) and edge computing have attracted much attention and appeal for a variety of real-time applications. IoT and edge computing applications operate in real time, necessitating a high data transmission capacity for day-to-day operations.

An efficient framework that is both energy efficient and simple to install in any system is required for use in real-world systems. One of the most significant difficulties for lightweight real-time applications is power usage. Current systems rely on bus-based interconnect architecture approaches, which are inefficient and frequently waste a lot of power, as well as creating performance bottlenecks [1]. A new on-chip connectivity design has been introduced to overcome these issues.

Network on chips (NoCs) have already made their way into signal/image processing and multiprocessor systems. Modern systems need high-speed data transmission rates as well as parallel communication capability. Accelerators in the form of a network provide faster deployment and enhanced algorithms for these devices [2]. These systems feature routers and

DOI: 10.1201/9781003415466-9

links that are responsible for linking and controlling data transit between distinct intellectual properties (IPs). This data routing is accomplished using data packets, which include the address of the recipient router as part of the data payload.

The hardware implementation of the NoC router is a challenging and crucial subsystem for any computing application. Implementation on actual electronics components such as GPUs, FPGAs, or ASICs is referred to as hardware-level implementation. These systems are critical for lightweight applications since they can be implemented on FPGA with scaling support for practically any IP core system with minimum changes. A lightweight application is one that uses a minimal amount of system resource such as memory and CPU. The implementation on FPGAs is an excellent example of such systems.

This work proposes a hardware implementation of routing computation for IoT-edge computing applications. As shown in Figure 9.1, the NoC connection proposed in this work is based on torus topology and employs XY routing. With Xilinx Vivado, the implementation is based on the Verilog hardware description language (HDL). There are two sides to the router: front and back. The data lines are managed by the front side, while the enable signals are managed by the back side. The router is built on multiplexers with cross bars. This NoC technology allows data to be sent in the form of packets with address information attached.

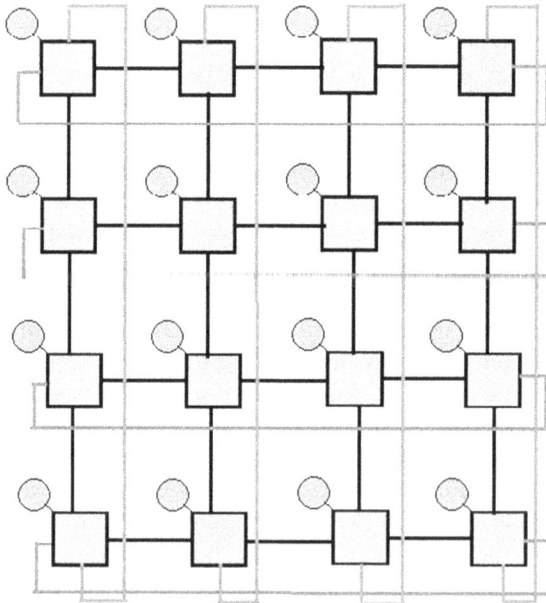

Figure 9.1 4 × 4 2D Torus NoC architecture.

9.2 LITERATURE SURVEY

A lot of research on NoC architecture and routing algorithms is being done worldwide. This section compares some existing research (see Table 9.1). In [3] an open-source 2D-NoC infrastructure for hardware acceleration using torus topology is proposed. Routing techniques for 2D-NoC using wrapper and system development are introduced in [2]. NoC interconnect is explored for multi-core image-processing system and its hardware-level prototype implemented on the Virtex II FPGA [4]. An NoC synthesis flow for a customized domain-specific multiprocessor system-on-chip (SoC) is discussed in [5]. This system can be scaled at edge computing level, connecting different processing elements (PEs) to each other using NoC interconnects.

The need for low power consumption and high-performance systems has encouraged researchers to implement many computing systems such as graphic processing units (GPUs), digital signal processors (DSPs), and various other IPs on a single system (usually referred to as SoC). The interconnection poses several challenges as the number of connecting blocks increases. Traditionally, shared bus architecture was used due to its simplicity and low-cost implementation, but it has many limitations such as wire delay and no support for parallel communication. For example, at a specific instance only one IP was able to control the bus using control signals, and all other IPs had to wait for one process to be completed. This increased latency, and low data internal speeds were achieved. Moreover, the performance was untenable in the case of a large number of processing elements. Network on chip was proposed to overcome this limitation.

The NoC approach helps to connect IPs, supported by parallel data communication, in a scalable way. Each IP is connected to a router, resulting in formation of a huge network. The IP could be the different components of an edge computing system or an IoT system or any other possible processing element. The main player inside the NoC is the router, which manages the link sending the data packets to the specified location. This work introduces the hardware-level implementation of a 2D network-on-chip architecture based on torus topology and XY routing.

Table 9.1 Comparison table

Ref.	Routing Scheme	Setup	Outcome	Remarks
[3]	Torus	Xilinx Virtex7	Low resource High clock performance	Cloud-based platforms
[4]	Mesh	Xilinx Virtex II	High performance for image processing	Real-time image-processing systems
[5]	Mesh	SystemC	NetChip	Video, network processor applications

9.3 METHODOLOGY

Figure 9.1 shows the basic infrastructure idea of a 4×4 2D NoC. Here the green box represents the router and the pink box represents the IP core. The way by which the routers are connected is called topology. There are various NoC topologies such as mesh, ring, torus and star. This work focuses on unidirectional torus topology due to its lightweight nature and potential for high clock performance [6]. Torus topology is just like mesh topology, the only difference being that it is cyclic. This means that the end element on the axis is connected to the starting element of that axis. As shown in Figure 9.2, the data size is chosen to be 8 bits and X, Y coordinates as 4 bits each, resulting in a total packet size of 20 bits; however, the data size is reconfigurable.

9.3.1 Router

The router is designed on two sides of the die for better understanding. The front side handles and shows all the data transmission routing, whereas the back side manages/controls all the enable signals for each data direction signal. Figure 9.3 shows the front side and Figure 9.4 the back side.

- **Front side:** The front side of the router manages incoming data and routes it to outgoing data based on the data address and router address. The router can receive data from the left and down directions as well

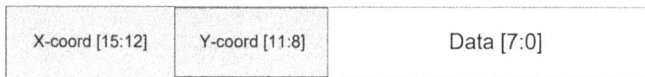

X-coord [15:12]	Y-coord [11:8]	Data [7:0]

Figure 9.2 Data format for payload and headers.

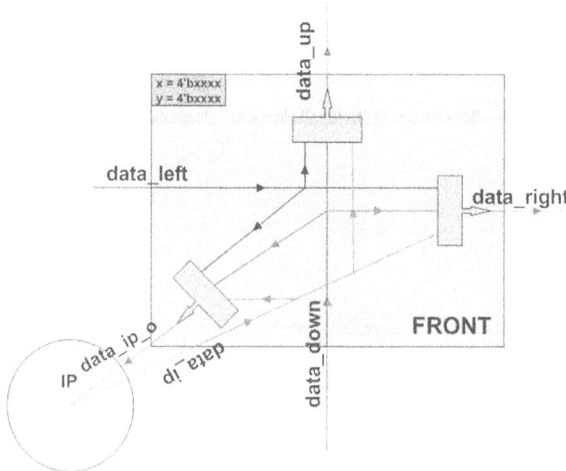

Figure 9.3 Router architecture front side.

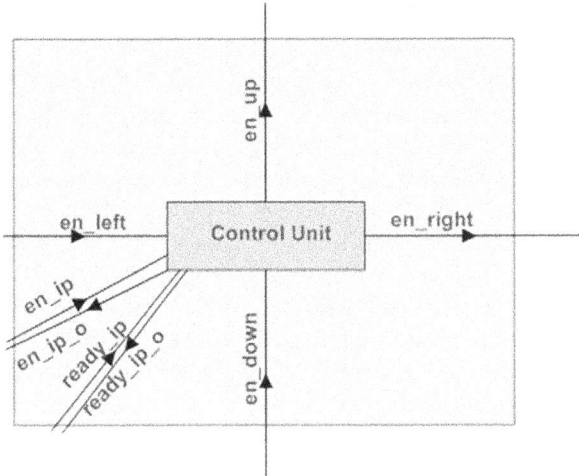

Figure 9.4 Router architecture back side.

as the IP core, and it can also send data in the right and up directions as well as the IP core. This routing is implemented using cross-bars multiplexers. The X coordinate is contained in *data_direction [15:12]*, Y coordinate in *data_direction [11:8]*; this information works as the select lines of the multiplexer. Due to the possibility that data packets could be available from the all the inlets at once, preferences are given according to the direction (see Section 9.3.2). There is also the possibility that data can be pushed backwards when all the routes are busy at a particular moment. It is assumed that the IP core has enough memory/buffer embedded in it for this purpose.

- **Back side:** The back side of the router manages all the control and enable signals. These signals basically establish if the data from a particular direction is valid or not. Each of the directions have one enable signal, whereas the IP core has two enable signals, one of which is for input data and the other for output data. For example, the *data_left* is only routed to some directions whenever *en_left* is high. There is also one internal signal, *ready_ip*, which determines if the IP core is ready to receive the data or not.
- **IP core:** The IP core is slightly different from the other direction lines. The IP core can even communicate with itself when needed; hence there is two-way data communication between the IP core and the two enable signals.

9.3.2 Routing

This section discusses the methodology by which many routers of the system will communicate with each other. As mentioned previously, this routing is implemented on the torus topology which uses XY routing. Each router of the system has its own address on the basis of its X and Y coordinates. In the case of multiple data arriving at the same time, the MUX handling the IP core is set to have the highest priority of the data coming from that specific IP core in order to achieve the minimum possible latency. Similarly, the MUX handling the up direction gives maximum priority to the data packets coming from the down direction, and MUX handling the right direction prioritizes data coming from left. This can be the main cause of a misrouting instance, and can result in slightly higher latency than usual. In such cases performance is reduced.

When the router receives the data packets, it compares the X–Y coordinates of the router to those of the coordinates held in the packet. If the address matches, the data is forwarded to the IP core, otherwise decisions are made, such as sending the data in the Y, i.e., up, direction if only the X coordinate matches, and in the X, i.e., right, direction if only the Y coordinate is matched.

9.3.3 Pseudo code

The pseudo code for the enable interconnects is as follows:

```
1. BEGIN
2. IF(data_left_x_coord == x_router_coord) and (data_left_
   y_coord != y_router_coord) and (data_left is valid))
3.          Set LeftToUp HIGH
4. IF(data_left_x_coord != x_router_coord) and (data_left_
   y_coord == y_router_coord) and (data_left is valid))
5.          Set LeftToRight HIGH
6. IF(data_left_x_coord == x_router_coord) and (data_left_
   y_coord == y_router_coord) and (data_left is valid))
7.          Set LeftToIP HIGH
8. IF(data_down_x_coord == x_router_coord) and (data_down_
   y_coord != y_router_coord) and (data_down is valid))
9.          Set DownToUp HIGH
10. IF(data_down_x_coord != x_router_coord) and (data_down_
    y_coord == y_router_coord) and (data_down is valid))
11.          Set DownToRight HIGH
12. IF(data_down_x_coord == x_router_coord) and (data_down_
    y_coord == y_router_coord) and (data_down is valid))
13.          Set DownToIP HIGH
14. IF(data_IP_x_coord == x_router_coord) and (data_IP_y_
    coord != y_router_coord) and (data_ip is valid) and
    (IP is ready))
15.          Set IPToUp HIGH
16. IF(data_IP_x_coord != x_router_coord) and (data_IP_y_
    coord == y_router_coord) and (data_ip is valid) and
    (IP is ready))
17.          Set IPToRight HIGH
```

```
18. IF(data_IP_x_coord == x_router_coord) and (data_IP_y_
    coord == y_router_coord) and (data_ip is valid) and
    (IP is ready))
19.         Set IPToIP HIGH
20. END
```

The pseudo code for the data flow is as follows:

```
1.  BEGIN
2.      case(data_direction):    //parameters not mentioned
                                    are don't care
3.      (DownToRight):              data_right <= data_
                                                    down
4.      (LeftToRight):             data_right <= data_
                                                    left
5.      (IPToRight):               data_right <= data_
                                                    ip
6.      (LeftToUp and DownToUp):   data_right <= data_
                                                    left
7.      (LeftToUp and IPToUp):     data_right <= data_
                                                    ip
8.      (IPToUp and DownToUp):     data_right <= data_
                                                    ip
9.      (LeftToIP and IPtoIP):     data_right <= data_
                                                    left
10.     (LeftToIP and DownToIP):   data_right <= data_
                                                    left
11.     (IPToIP and DownToIP):     data_right <= data_
                                                    down
12.     (LeftToIP and (IP not ready)): data_right <= data_
                                                    left
13.     (IPtoIP and (IP not ready)):   data_right <= data_
                                                    ip
14. (LeftToUp and DownToIP and (IP not ready)):data_right
                                            <= data_
                                                    down
15. (IPToUp and DownToIP and (IP not ready)):  data_right
                                            <= data_
                                                    down
16.     (DownToUp):                data_up   <= data_
                                                    down
17.     (LeftToUp):                data_up   <= data_
                                                    left
18.     (IPToUp):                  data_up   <= data_
                                                    ip
19.     (DownToRight and LeftToRight):  data_up <= data_
                                                    left
20.     (DownToRight and IPToRight):    data_up <= data_
                                                    IP
```

```
21.(DownToRight and LeftToIP and (IP not ready)):data_up
                                        <= data_
                                           left
22.(DownToRight and IPToIP and (IP not ready)):  data_up
                                        <= data_
                                           IP
23.    (LeftToRight and IPToRight):       data_up   <=
                                           data_
                                           IP
24.(LeftToRight and IPToIP and (IP not ready)):  data_up
                                        <= data_
                                           IP
25.(LeftToIP and IPToRight and (IP not ready)):  data_up
                                        <= data_
                                           left
26.(LeftToIP and IPToIP and (IP not ready)):     data_up
                                        <= data_
                                           IP
27.(DownToIP and IPToIP and (IP not ready)):     data_up
                                        <= data_
                                           IP
28.    (DownToIP and (IP not ready)):     data_up   <=
                                           data_
                                           down
29.    (IPtoIP and (IP ready))  :         output_data_
                                           ip<= data_ip

30.    (DownToIP and (IP ready)):         output_data_
                                           ip<= data_down

31.    (LeftToIP and (IP ready)):         output_data_
                                           ip<= data_left
32. end case
33. END
```

9.4 RESULT

For the simulation and estimation of area and power overheads of the NoC router, Xilinx Vivado version 2018.2 is used. The area and power factors of a design play a very important role in determining the efficiency and latency of the designed chip. Xilinx Vivado was used to generate the area and power overhead reports.

The router for the hardware-level implementation is designed on the FPGA level as it is lightweight, requiring low system resources. It was later implemented on the Xilinx Vivado tool. The functional verification of the routing computation unit is performed using Vivado Simulator. The waveform for the functional verification is presented and some simulation cases are shown in Figure 9.5. The first cycle shows the system reset. The next

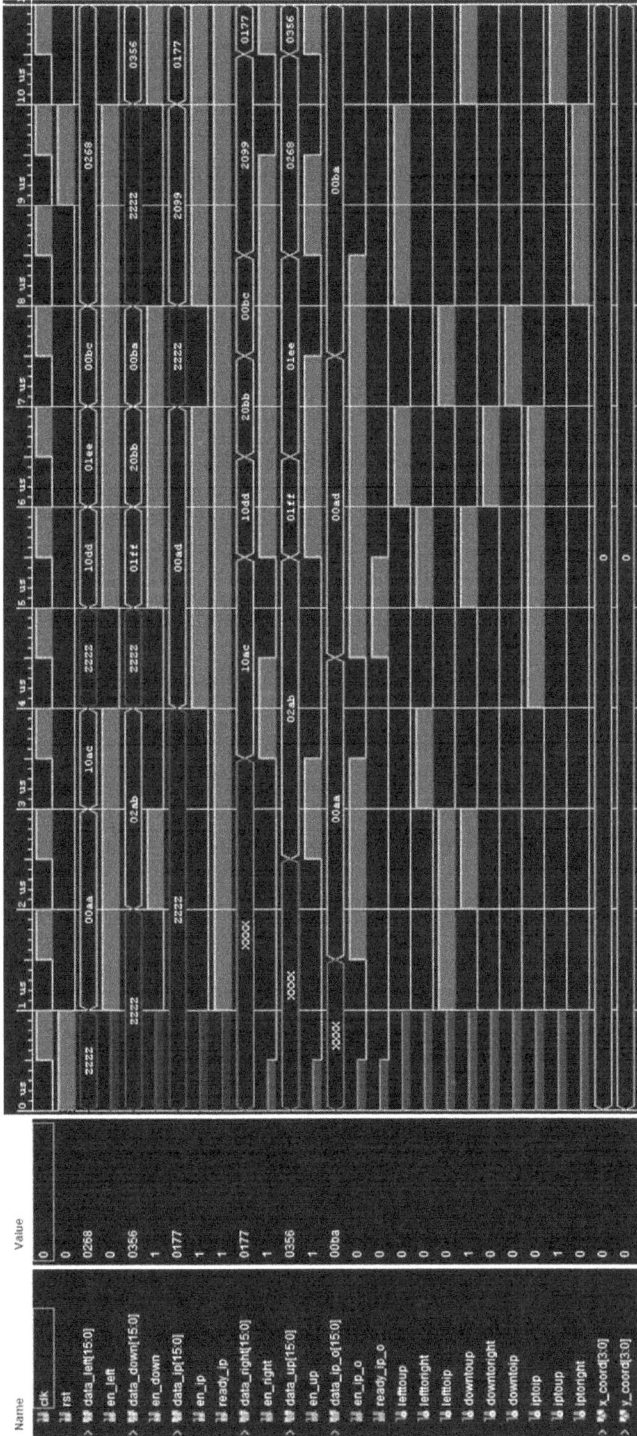

Figure 9.5 Simulation timing diagram.

cycle input sent data from the left to the IP. For that, X and Y coordinates were sent equivalent to the router and data sent was *AA* in hexadecimal. For the next cycle, data was sent from the left to the IP and down to up simultaneously. The next cycle shows data sent from left to right and then from IP to IP. For the next cycle, data was sent on three channels simultaneously, i.e., left to right, down to up, and IP to IP. The following cycles are simulated in the same way. In the tenth cycle, a reset signal was again given; this resets all the output enable signals, meaning that data becomes invalid when reset is set high. In the next cycle, data was sent from IP to up, and down to up, simultaneously. Since data is coming from two ports to the same direction, priority settings are used (see Section 9.3.2). In this case, data is routed from down to up and IP to right.

The ports used in the design are:

    i. *Clk*: Input; takes the system clock signal for functioning.
    ii. *Rst*: Input; resets the whole system when set as high.
    iii. *data_left*: 16bits input; carries the data from the left to the router.
    iv. *En_left*: Input; determines if the data coming from the left is valid or not.
    v. *Data_down*: 16bits input; carries the data from the bottom to the router.
    vi. *En_down*: Input; determines if the data coming from the bottom is valid or not.
    vii. *data_ip*: 16bits input; carries the data from the IP to the router.
    viii. *En_ip*: Input; determines if the data coming from the IP is valid or not.
    ix. *Ready_ip*: Input; determines if the IP is ready to receive data or not.
    x. *Data_right*: 16bits output; carries the data from the router to the right.
    xi. *En_right*: Output; determines if the data going to the right is valid or not.
    xii. *Data_up*: 16bits output; carries the data from the router to the right.
    xiii. *En_up*: Output; determines if the data going to the right is valid or not.
    xiv. *Data_ip_o*: 16bits output; carries the data from the router to the right.
    xv. *En_ip_o*: Output; determines if the data going to the right is valid or not.
    xvi. *Ready_ip_o*: Output; determines if the IP will be ready to receive data in the next iteration.

The wires used were named simply according to which direction they are propagating, in the *DirectionToDirection* format, namely *lefttoup, lefttoright, lefttoip, downtoup, downtoright, downtoip, iptoip, iptoup,* and *iptoright.*

This project was set up and created on Xilinx Vivado 2018.2. The RTL design is shown in Figure 9.6. It was synthesized and the synthesis result is

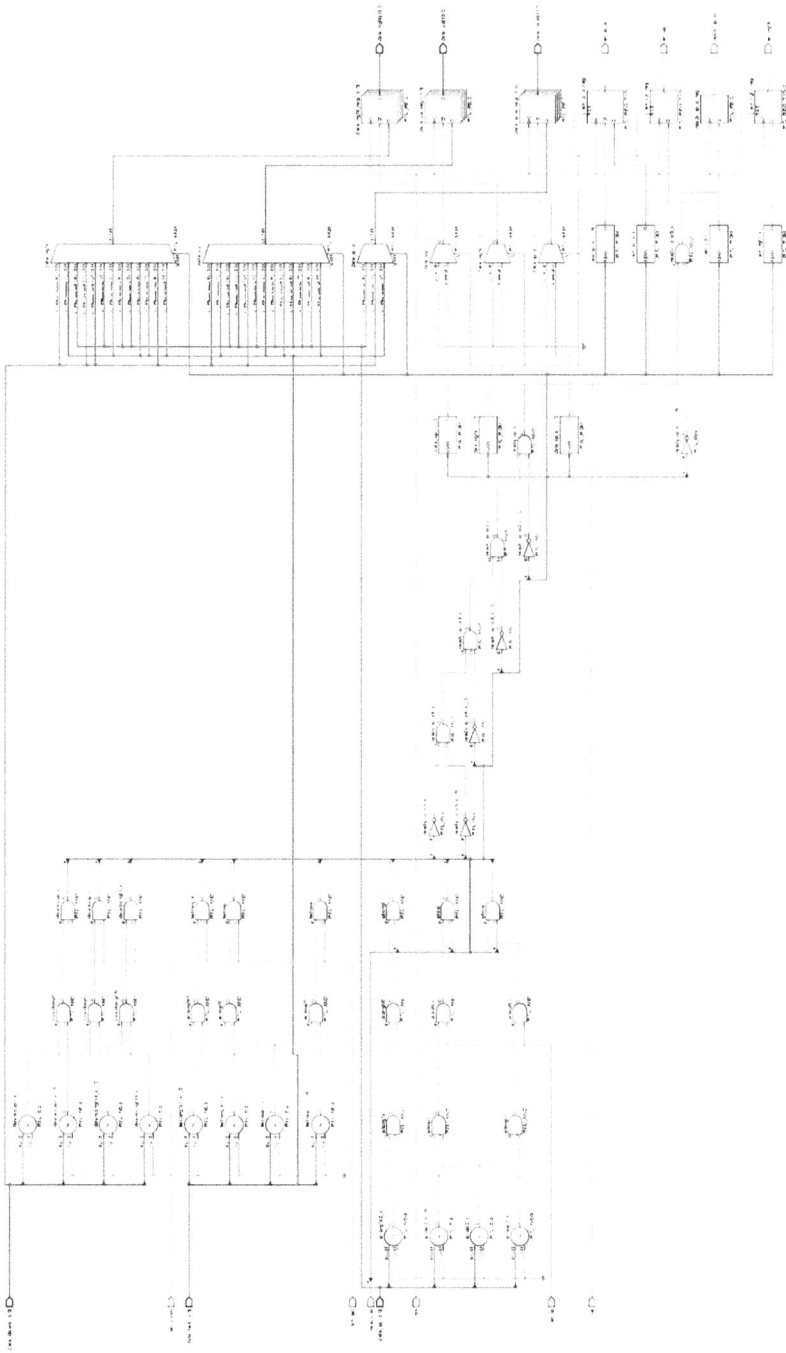

Figure 9.6 RTL design.

```
Start RTL Hierarchical Component Statistics
---------------------------------------------
Hierarchical RTL Component report
Module nocrouter
Detailed RTL Component Info :
+---Registers :
                    16 Bit     Registers := 3
                     1 Bit     Registers := 4
+---Muxes :
            31 Input    16 Bit       Muxes := 3
            31 Input     1 Bit       Muxes := 9
```

Figure 9.7 Synthesis report.

Table 9.2 Area report

	Area report		
Number of ports	106		
Number of nets	266	Combinational	369.52538
Number of cells	212	Buf/Inv area	36.85088
Number of combinational cells	160	Noncombinational area	290.74075
Number of sequential cells	44	Macro/black box area	0
Number of macros/black boxes	0	Net interconnect area	142.08489
Number of buf/inv	29	Total cell area	660.26612
Number of references	20	Total area	802.35102

Table 9.3 Power report

Power group attrs	Switching power	Leakage power	Total power
Sequential	3.06E-04	6.20E+06	6.1965
Combinational	0.2327	1.12E+07	11.4275
Total (uW)	0.233	17.391	17.624

shown in Figure 9.7. The area and power report that are shown in Tables 9.2 and 9.3 are generated using Vivado and some open-source tools.

9.5 CONCLUSION

High-performance computing applications require high-speed interconnects that do not throttle back their performance in real life. The traditional bus architecture system does not fulfill the need of modern-day computing systems such as IoT and edge computing applications which require high-speed internal connection. This work has described the hardware-level implementation of network-on-chip (NoC) systems designed to overcome these bottlenecks in IoT and edge computing applications. There is still lot of scope for improvement in performance of NoC based systems. Some existing methodology includes implementation of flexible router architecture [7].

REFERENCES

[1] Ofori-Attah, Emmanuel, Bhebhe, Washington, and Opoku Agyeman, Michael. (2017). Architectural techniques for improving the power consumption of NoC-Based CMPs: A case study of cache and network layer. *Journal of Low Power Electronics and Applications*, vol. 7, p. 14. 10.3390/jlpea7020014.

[2] Kumar, Shashi et al., A network on chip architecture and design methodology, *Proceedings IEEE Computer Society Annual Symposium on VLSI. New Paradigms for VLSI Systems Design*, Pittsburgh, PA, USA. ISVLSI 2002, 2002, pp. 117–124, doi: 10.1109/ISVLSI.2002.1016885.

[3] Reddy, Kuladeep Sai and Kizheppatt Vipin, OpenNoC: An open-source NoC infrastructure for FPGA-based hardware acceleration, *IEEE Embedded Systems Letters*, vol. 11, no. 4, pp. 123–126, Dec. 2019, doi: 10.1109/LES.2019. 2905019.

[4] Joshi, Jonathan, Karandikar, Kedar, Bade, Sharad, Bodke, Mandar, Adyanthaya, Rohan, and Ahirwal, Balkrishan. "Multi-core image processing system using network on chip interconnect," in *2007 50th Midwest Symposium on Circuits and Systems*, 2007, pp. 1257–1260. 10.1109/MWSCAS.2007.4488781.

[5] Bertozzi, Davide et al., NoC synthesis flow for customized domain specific multiprocessor systems-on-chip, *IEEE Transactions on Parallel and Distributed Systems*, vol. 16, no. 2, pp. 113–129, Feb. 2005, doi: 10.1109/TPDS.2005.22.

[6] Kapre, Nachiket and Jan Gray, "Hoplite: Building austere overlay NoCs for FPGAs," in *2015 25th International Conference on Field Programmable Logic and Applications (FPL)*, London, UK, 2015, pp. 1–8, doi: 10.1109/FPL.2015.7293956.

[7] Hassan, Hossam, Ragab, Mohamed, Sayed, Mohammed, and Goulart, Victor. "Hardware implementation and evaluation of the flexible router architecture for NoCs," in *2013 IEEE 20th International Conference on Electronics, Circuits, and Systems*, Abu Dhabi, UAE, ICECS 2013, 2013, pp. 621–624. 10.1109/ICECS.2013.6815491.

Chapter 10

Artificial intelligence-based techniques for operations research and optimization

Katakam Venkata Seetharam

Sreenidhi Institute of Science and Technology, Hyderabad, India

CONTENTS

10.1 INTRODUCTION

The Operations Research Society of America defines operations research (OR) as the science of "deciding how to best design and operate man–machine systems" within scarce resources. It is generally accepted that the OR discipline had its origin during World War II, when the teams of scientists studying deployment of military hardware tried to arrive at the best options to minimize losses and maximize returns. OR now refers to the scientific base employed to obtain optimal and effective decisions in order to maximize a certain benefit or minimize a certain cost for a variety of

DOI: 10.1201/9781003415466-10

resource allocation problems such as plant location, inventory control, production scheduling, portfolio selection, etc.

Essentially, OR involves taking decisions to achieve optimal outcomes or rewards working with limited resources. In this context, OR considers interactions between the sub-systems or components of the overall system and formulates a model to achieve the optimal outcome. Most OR problems can be categorized as follows:

1. *Resource allocation problems*. When a limited resource has to be allocated among competing candidates we formulate a model that takes into account rewards accrued due to a unit allocation to different candidates. The allocations are 'constrained' in such a way that the total allocations should be within the given resource and ensure any specified minimum allocations to different candidates. These problems materialize as linear programming models, when the objective function and constraints are linear, or nonlinear programming models, when all or some of the functions involved are nonlinear.

2. *Multistage sequential decision making*. In these systems the decisions are spread across multiple sequential stages and separate transfer functions or continuity equations maintain the continuity of the system states across the stages. The goal is to maximize the benefit accruing from the total stages of the process and these problems form dynamic programming models.

3. *Queueing models*. These models deal with service providers (servers) and service takers (customers) and aim to optimize the overall cost involving cost of establishing the servers and cost of waiting experienced by customers – in terms of the level of satisfaction. Before arriving at an optimum solution, these models need extensive analysis to determine certain performance parameters such as average waiting time, average service time, length of the queue at any given time, etc., where each parameter follows a particular probability distribution. As there is a tradeoff between the cost of providing a server and the cost of satisfying a customer, these problems involve multi-objective optimization models.

4. *Inventory models*. These models aim to maintain minimum inventories in line with sales without disrupting sales due to a shortage. As an inventory implies locked-up capital, minimizing the inventory means maximizing the productivity of the capital employed. In this case also, extensive analysis is required to establish demand patterns, lead times between order and receipt, estimates of the cost of disruption to sales, etc. These problems usually involve nonlinear terms in the objective functions and constraints, forming nonlinear programming or dynamic programming models.

5. *Network models*. A network consists of a set of nodes and the connections among these nodes, called arcs. Finding the shortest distance

between two cities from the various possible routes, and the minimum-cost distribution of goods from manufacturing plants to warehouses to retail shops, are examples of network problems. The 'traveling sales-man problem' (TSP), where a salesperson needs to start at one node and cover all the other nodes, only once, with minimum total distance traveled, is also a network problem. These problems usually present considerable computational challenges due to the huge number of combinations. Linear programming can be applied to small-sized problems as the formulation results in a large number of constraints.

After the OR problem is formulated, with an objective function, and the relationship between the variables and constraints, the final step is to find the optimal solution among the feasible solutions, using various optimization techniques. As the solution of OR problems consists in employing suitable optimization techniques, we conveniently group OR problems with optimization problems as OR&O problems.

As discussed above, an OR problem finally materializes as an optimization problem in the form of a linear, nonlinear, or dynamic programming problem to be solved to obtain the optimum decisions. However, only linear programming problems can be solved efficiently, and both nonlinear and dynamic programming problems present considerable challenges in terms of computational and memory loads. In this context, employing artificial intelligence (AI)-based techniques that use certain learning and knowledge-based algorithms to obtain practical and useful solutions for OR&O problems with reasonable computational resources has gained widespread acceptance. We present a general OR&O problem formulation in Section 10.2, AI-based solution techniques in Section 10.3, and conclusions in Section 10.4.

10.2 PROBLEM FORMULATION

The OR&O formulation involves an objective function, involving several variables, to be minimized or maximized satisfying the given constraints. When stochasticity is introduced by random processes, additional constraints in the form of reliability of the system performance will become necessary.

$$\text{Maximize} f(X,T) \atop X \tag{10.1}$$

subject to the constraints set

$$G(X,T) \le 0 \tag{10.2}$$

and variable bounds

$$X_{\min} \le X \le X_{\max} \quad T_{\min} \le T \le T_{\max} \tag{10.3}$$

where $X \in R^n$, $n \in Z^*$ is a vector of decision variables; objective function, $f(\ ): X \to R$ is a vector function; constraints function set, $G(\ ): X \to R^g$ is a set of vector functions; $T \in R^m$, $m \in Z^*$ is a state variables vector; b is a vector of constant terms; $Z_{\min}$ and $Z_{\max}$ denote lower and upper bounds for any Z. In certain types of OR&O problems state variables, representing states of the system, influence the objective function. For example, in hydropower production the power generated depends on flow rate (decision variable), and elevation of the water, which is linked to the current storage (state variable) of the reservoir. Further, the system state, representing the storage of the reservoir, will have a maximum bound, which is the capacity of the reservoir. The state variables are not independent variables, and the relation between state and decision variables is given by a transfer function (equality constraint).

$$t(X,T) = 0 \tag{10.4}$$

However, for most of the optimization methods the above formulation is conveniently expressed only in terms of decision variables as (retaining the notation of the above formulation):

$$\begin{array}{c} \text{Maximize}\, f(X) \\ X \end{array} \tag{10.5}$$

subject to

$$G(X) \le 0 \tag{10.6}$$

$$H(X) = 0 \tag{10.7}$$

$$X_{\min} \le X \le X_{\max} \tag{10.8}$$

When all the functions, f, G, H, and t, are linear, we say the problem is linear, and efficient linear programming methods can be applied to find the analytical solution for the optimum. If any of the above functions is nonlinear, we have a nonlinear problem at hand and the solution depends on the characteristics of functions such as continuity, multimodality, etc. For a set of functions known as convex functions which have a single peak or mode in the entire search domain, we usually have analytical solutions for the optimum. Though nonlinear programming methods can work with continuous functions, the existence of constraints and the multimodal nature of the functions result in finding only a local optimum.

When there are no constraints ($G(\ )$: $X \to null$), we say the problem is an unconstrained optimization problem, which is less complicated than a constrained optimization problem where constraints introduce complexities in the search process. In a constrained problem, any X that satisfies the constraints and variable bounds is called a feasible solution. Therefore, the optimum needs to be found within the feasible solution set called the 'decision space'.

When the objective function, $f(\ )$: $X \to R^o$ where $O > 1$ we have a multi-objective problem and cannot have a single optimum value, as, in general, the objectives would be conflicting. Instead, here the solution space consists of pareto-optimal or nondominated solutions. A solution is called pareto-optimal or nondominated if improving any one objective results in degrading at least one other objective. Therefore, first we need to arrive at the pareto-optimal solutions from which a final solution can be taken, based on the preferences and trade-offs available with the decision maker.

When any input to the system or the function is random, we have the stochastic optimization problem. The stochastic nature of the variables and functions introduces additional dimensionality to the problem, greatly increasing the search space.

The methods used to solve optimization problems depend on the above problem formulation factors. For linear problems we have efficient simplex-based linear programming (LP) methods to find the optimum. For nonlinear problems, nonlinear programming (NLP) methods are available but the efficiency of these methods depends on the nature of the functions in the formulation. For multistage problem formulations with arbitrary and discontinuous objective and constraint functions, an efficient recursive method called dynamic programming (DP) is available. Moreover, DP formulation is suitable for considering random processes, as in stochastic dynamic programming (SDP). However, as DP works in stages, keeping track of all the possible 'states' of the system, the method is burdened by the number of states in the problem, which is known as the 'curse of dimensionality'. The computations involved increase exponentially with the number of states in the system, limiting its usefulness for complex real-life problems.

When available analytical methods are not suitable for the problem at hand, we resort to trial-and-error-based methods such as simulation. Simulation is a systematic technique of obtaining response from a system, which is represented in all its characteristics by a suitable mathematical model, for a given set of inputs. The inputs consist of exogenous inputs, i.e., the inputs to the system from outside, and values for decision parameters which are part of the mathematical model of the system. From the set of all responses obtained by varying the parameters systematically, conclusions can be drawn as to what the decision parameters should be that suit the objective of the project. Though simulation is a useful technique, we still need to narrow down the search space, since considering all possible combinations of the decision parameters would create a prohibitively large computational load.

Therefore, in practice, simulation incorporates a sub-optimization scheme or a screening model to narrow down the vast parameter space to promising regions that can be explored with simulation.

Heuristic methods employ direct search techniques to locate promising regions step by step, leading to approximate solutions. When artificial intelligence (AI) is employed in these search techniques, known as metaheuristics, powerful algorithms emerge, promising solutions that are very near to the optimum, with minimum computational effort. As no comprehensive and general technique exists that can address the most critical and practical aspects of OR&O problems encompassing discontinuous, non-differentiable, multimodal cases, these AI-based methods have become useful for finding near-optimal results efficiently.

10.3 AI-BASED METAHEURISTICS/OPTIMIZATION METHODS

AI is generally understood as the extension of the human intelligence process to computational systems, involving learning from the given data and acquiring knowledge, from which the system then predicts the behavior or output, given the inputs. Broadly, including powerful ideas of natural evolution with AI, we consider AI-based metaheuristics under the following three categories:

1. Evolutionary algorithms
2. Other nature-inspired algorithms
3. Artificial neural networks.

10.3.1 Evolutionary algorithms

Evolutionary algorithms (EAs) are population-based parallel search algorithms which apply natural evolutionary processes on an initial set of working solutions with the aim of obtaining an improved set of solutions in each iteration. Evolutionary algorithms ensure a collective learning process within a population of individuals, each of which represents a search point in the feasible search domain. In each iteration or generation, the population of individuals are randomly varied and selected. Selection retains only the individuals having better fitness values, discarding the inferior ones. This variation of solutions and selection pressure carried out over generations leads to a trail of improved solutions. At the end of some satisfactory criteria, the solution which has the largest fitness value is considered the optimum solution.

Here, we consider under EAs, for OR&O problems, evolutionary strategies (ES) [1, 2], genetic algorithms (GA) [3, 4], and variants based on these methods such as evolutionary programming (EP) [5], and differential evolution (DE) [6]. As genetic algorithms is the most popular and widely studied

[7] of the EAs, we shall go into the detailed workings of GA followed by a brief explanation of other EA methods.

10.3.1.1 Genetic algorithms (GA)

The idea of GA, or in general of EA, is that if each point in the search region is assigned a fitness, according to a quantifiable objective function, then the optimum, which has the greatest fitness associated with it, can be approached by starting with a randomly constructed pool of solutions (represented in coded form), and exploring it successively with a set of genetic operators, forming trajectories of solutions which come closer and closer to the optimum.

Following the terminology of natural evolution, each point in the feasible solution space can be thought of as an individual, and a population of individuals comprises a set of points in the feasible solution space. The representation of a (feasible) solution (or individual) consists of an ensemble of all the corresponding decision variables values encoded in binary or real mode, called chromosome or string.

GA works by successively exploring (or evolving) a randomly initialized population (where each bit of a string assumes either '0' or '1' with (equal) probability 0.5), of size N (i.e., N number of strings), according to the laws of natural evolution – survival and propagation of the fittest – implemented by genetic operators, selection, crossover, and mutation. Selection gathers the best strings, in terms of fitness, from the existing population, on which crossover and mutation work by effecting a different combination of constituent bits in the strings, to produce a new population. The process of obtaining a new population from the existing population – termed a generation – is carried out – with average fitness improved in each passing generation – until some satisfactory criteria are met. Randomness is introduced at each level of the evolution to allow diverse individuals to thrive, which may confer beneficial advantage to the population due to the better adaptation of these individuals in future (generations).

As a string comprises a concatenation of substrings, each of which is coding for a distinct decision variable, the fitness of the string can be attributed to certain lengths of substrings, called substructures or schemata [3] that code for a favorable composition of some or all of the decision variables. A schemata stands for some combination of substrings of certain length where some specified number of bit positions are fixed (with predetermined values, '1' or '0'), and the remaining bit positions are flexible (can take any '1' or '0'). The defining length of the schemata is the string length between the first and last fixed positions, and the order of the schemata is the number of fixed positions in it. For example, the schemata '1***011*0', having 5 fixed positions and 4 flexible positions denoted by '*', is of order 5 with defining length 8. Holland's schema theorem or building-block hypothesis [4] predicts that small defining length and lower-order highly fit schemata increase exponentially in the population through selection mechanism and nondisruptive crossover and mutation. Implicit in the above theorem is that the schemata keep getting better under crossover and mutation operators. Since only the better strings, produced from crossover and mutation, are allowed to survive (under selection pressure) the (average) fitness of the strings increases monotonously.

GA algorithm:

initialize *population*	initialize a population of individuals each representing a feasible solution
evaluate *fitness*	fitness is equal or proportional to the objective function
generation $g = 0$	
while (! *end condition*)	based on incremental convergence value, etc.
select individuals for next generation	selection is based on the fitness value
perform *crossover*	recombine the portions of the coding regions among the pairs of individuals
perform *mutation*	mutate the coding portions randomly
evaluate *fitness*	one generation is completed
generation $g = g+1$	
solution = *the best individual*	

Coding: The first step in the GA is coding the decision variables in the problem. Coding, or the actual representation of the ensemble of decision variables making a solution point, is done either in binary mode or in real-number mode, forming a string of binary or real numbers. The binary coding is generally suitable for all problems and offers maximum parallelism, since it offers the maximum number of schemata per one bit of information [3]. However, in dealing with problems with a large number of decision variables, the binary string may become too large, which may cause easy disruption of competent solutions while evolving. In real coding, since the string length is small, the risk of disruption of the competent string is small [8]. Ensuring genotype (coding space) and phenotype (solution space) are contiguous, gray coding considers a suitable mapping so that contiguous strings differ by unit value in their decoded values. For a 2-bit length, strings '00,01,11, and 10' constitute four successive contiguous strings with decoded values of 0,1,2, and 3 respectively. For example, let r_1, x_1, and π_1 be decision variables with feasible ranges $0 \leq r_1 \leq 8$, $0 \leq x_1 \leq 16$, and $0 \leq \pi_1 \leq 90$ respectively. Since a binary string of length n can represent a maximum value of (2^n-1), the above variables can be represented using (sub) strings of length 4, 5 and 7 bits respectively for r_1, x_1, and π_1, covering their entire feasibility ranges, and the concatenation of all the substrings coding for different decision variables forms a full string (or chromosome) of that particular solution (or individual).

Selection: Strings are selected from the existing population such that the number of copies of each string is in some way proportional to its fitness. The selected strings undergo further genetic mechanisms to finally form a new generation. In the simplest deterministic selection method, each string gets copies according to the integer part in the expected number (of copies), found by computing $f_i / \bar{f}$ where f_i is fitness of the string i, and $\bar{f}$ is average fitness of the population of size n given by $\Sigma_i f_i / n$. The fraction parts are sorted out, and the remaining population is filled with strings from the top of the order.

Crossover: This operator effects recombination among different strings selected by a selection method and allows competent schemata present in different strings to aggregate. In the single-point crossover, a pair of strings exchange corresponding parts of their sequence across a randomly chosen point of the string (Figure 10.1). In uniform crossover, a uniformly distributed random string is generated which operates on a set of parent strings to create a child string taking a bit from a particular parent based on the parity of the corresponding bit of the random string. Though maximum recombination can be assured in uniform crossover, the survival rate of any competent schema will deteriorate. Crossover probability, i.e., given a pair of strings, what is the probability that the crossover has to be implemented, is an important parameter. During initial generations its value is usually kept high and during final stages of convergence it is kept low to preserve good solutions.

Mutation: Although selection and crossover are the main mechanisms on which GA depends to explore the search space systematically, it is possible that the entire population may have turned to have the same string in all its individuals at some point during the generations. In this situation the crossover operator breaks down so that it can no longer alter any bits of the string, since there is no way to introduce any new bits in the strings.

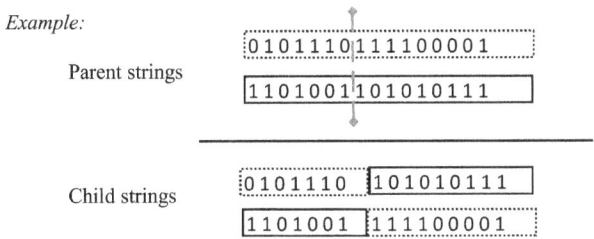

Example:

Parent strings

0101110 111100001

1101001 101010111

Child strings

0101110 101010111

1101001 111100001

Figure 10.1 Single-point crossover operation.

To tackle this problem, a mutation operator, which randomly flips the bits of the string (i.e., changing bit '1' to bit '0' and vice versa), is used in GA. Since its principal use is to introduce genetic diversity into the population, and mutation can potentially disrupt good strings, its probability is maintained low.

Example:

Given string:	0 1 0 1 1 1 0 1 0 1 0 1 0 1 1 1
Mutated string:	0 1 0 0 1 1 0 1 0 1 1 1 0 1 1 1
(4th and 11th bits are flipped)	

Once the mutation is done, one generation is completed and the new population will have greater average fitness than the previous one. With the new population the process repeats till a set criterion is reached or a predetermined number of generations are completed.

The constraints in GA are usually dealt with by penalizing the fitness of infeasible solutions, where a penalty function, in some way, awards a penalty according to the severity of the violation of the constraints. In the death penalty method, the fitness of the infeasible solution is made zero. Some methods explore the search space for strings which satisfy the constraints sequentially, one at a time, till all the constraints are satisfied. Some methods repair the infeasible strings, i.e., an infeasible string will be mapped into a nearest feasible string.

Niching: In the case of multimodal function space, all the dominant peaks in the search space are required to be identified. Holding on to close-magnitude peaks at the same time can also be advantageous in further searches. Since the usual GA operators predominantly favor a single peak, an additional operator, niching, is necessary to force the strings to share the peaks or niches in proportion to its magnitude. Niching helps maintain the diversity of the individuals, similar to nature which encourages stable populations in different niches. In one method niching can be achieved by scaling the fitness of a string inversely proportional to its nearness to other strings. The nearness can be computed in terms of hamming distance, obtained by adding up the (absolute) differences between corresponding alleles of two strings.

Nondominated Sorting Genetic Algorithm (NSGA): For multi-objective problems, as there is no single objective function, we need to find the best nondominated or pareto-optimal front, consisting of pareto-optimal solutions. NSGA, devised by Srinivas and Deb [9], evaluates the entire population and sorts them into a series of pareto-optimal fronts (Figure 10.2) beginning with the highest-ranked front. Each solution in the highest-ranked pareto-optimal front shares, as per a sharing function, an assigned (highest) fitness value, and the solution points in the next-ranked front will have a smaller assigned fitness value, and so on. The selections to the next generation are

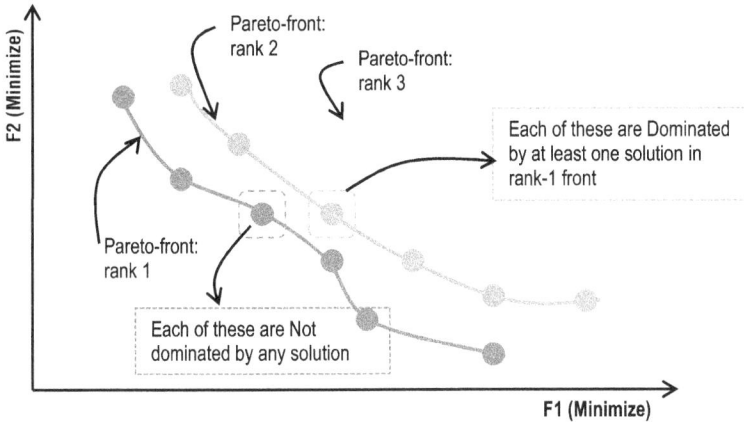

Figure 10.2 Pareto-optimal fronts.

carried out based on the fitness value, ensuring that the solutions in a front have an equal chance of being selected, maintaining the diversity of the solutions. After a series of generations, the population is expected to converge to the global or near-global pareto-optimal front. Deb et al. [10] improved the NSGA, which is computationally heavy ($O(MN^3)$, where M is the number of objectives and N is population size), with an elitist NSGA-II, where the computational burden is reduced (to $O(MN^2)$), and a crowding comparison feature for better fitness sharing.

NSGA-II algorithm [10]:

Generate a random population P_0 (*size N*)
Sort P_0 based on nondomination and assign ranks to each $p \in P_0$
Assign fitness to each p based on nondomination rank
Create offspring population Q_0 (*size N*) using normal *selection, crossover and mutation*

$t = 0$	Implementation of elitism
$R_t = P_t \cup Q_t$	Combine parent and offspring populations
$F = (F_1, F_2, ..)$	F is nondominated sorting of R_t
{ for each $p \in P$	***nondominated sorting procedure***
$S_p = \varnothing$;	S_p: set of solutions dominated by p
$n_p = 0$	n_p: number of solutions that dominate p

for each $q \in P$

if $(p \prec q)$ then $S_p = S_p \cup \{q\}$ if p dominates q add q
                                                                                            to S_p

else if $(q \prec p)$ $n_p = n_p + 1$

if $n_p = 0$ then $p_{rank} = 1$; $F_1 = F_1 \cup \{p\}$ p belongs to 'rank-1'
                                                                                            front

$i = 1$ initialize the front
                                                                                            counter

while $F_i \neq \varnothing$

$Q = \varnothing$ initialize next front

for each $p \in F_i$

for each $q \in S_p$

$n_q = n_q - 1$

if $n_q = 0$ then $q_{rank} = i+1$; $Q = Q \cup \{q\}$ q belongs to $(i+1)$ front

$i = i + 1$

$F_i = Q$

}

$P_{t+1} = \varnothing$ and $i = 1$

until $|P_{t+1}| + |F_i| \leq N$

crowding-distance assignment (F_i)   calculate crowding
                                                                                            distance in F_i

Crowding-distance assignment procedure

{ $S = |F_i|$ number of solutions in F_i

$F_i(s)_{distance} = 0, 0 \leq s \leq S$ initialize crowding
                                                                                            distance

for each objective m $F_i[s]_{distance}$ is cumulative
                                                                                            over all the objectives

$F_i = sort(F_i, m)$ sort F_i in ascending
                                                                                            order as per objective
                                                                                            value f_m

$F_i[1]_{distance} = F_i[S]_{distance} = \infty$ to make boundary
                                                                                            points always selected

for $s = 2$ to $(S-1)$

$F_i[s]_{distance} = F_i[s]_{distance} + (F_i[s+1,m] - F_i[s-1,m]) / (f_m^{max} - f_m^{min})$

}

$P_{t+1} = P_{t+1} \cup F_i$

$i = i + 1$ prepare the next counter
                                                                                            for inclusion

sort $(F_i, \prec)$ sort in descending
                                                                                            order using crowded-
                                                                                            comparison operator, $\prec$

Crowded-comparison operator ($\prec$) procedure

Say, i_{rank} = non domination rank of 'i'
and, $i_{distance}$ = crowding distance of 'i'
then, define $\prec$ as, $i \prec j$ if $(i_{rank} < j_{rank})$ nondominated rank
counts first

or $((i_{rank} = j_{rank})$ *and* $(i_{distance} > j_{distance}))$ preferring more spaced
candidates

$P_{t+1} = P_{t+1} \cup F_i[1: (N - |P_{t+1}|)]$ Choose the first
$(N - |P_{t+1}|)$ of F_i

$Q_{t+1} = $ *make new pop* (P_{t+1}) Create new population
with normal selection,
crossover, mutation
$t = t + 1$ increment the generation
counter

10.3.1.2 Evolution strategies (ES)

ES can be distinguished from GA by the use of a probability distribution-based mutation operation. Mutation is the primary exploration tool in ES. ES basically employs real values for the variables and an individual contains the set of the variables, objective function and a mutation strategy set. Recombination in ES varies from generating one individual from two parents to taking components from all the individuals to generate one individual. Selection is done based on the objective function from a large set of candidate solutions generated. The basic algorithm is as follows [11]:

ES algorithm:

$g = 0$ generation = 0
initialize:$P_p^{(0)} := \{(y_m^{(0)}, s_m^{(0)}, F(y_m^{(0)})), m = 1, ..., \mu\}$ y: variable set, s:
strategy set , F() is
objective function,
μ: parent popula-
tion size

for l =1 to λ begin
$M_l=$ *marriage* $(P_p^{(0)}, \rho)$ take 'ρ' individuals
$s_l:=s\_recombination$ (M_l) new s from
recombination

$y_l:=y\_recombination$ (M_l) new y from
recombination

$\check{s}_l:=s\_mutation$ (s_l) new s from mutation
$\check{y}_l:=y\_mutation$ $(y_l, \check{s}_l)$ new y from mutation
$\check{F}_l :=F(\check{y}_l)$
end

$P_o^{(g)} := \{(\breve{y}_l, \breve{s}_l, \breve{F}_l), l = 1, ..., \lambda\}$ make offspring
population size of λ

choose a *selection* type

$(\mu, \lambda) : P_o^{(g+1)} := selection \ (P_o^{(g)}, \mu)$ *select* from the
current set only

$(\mu + \lambda) : P_o^{(g+1)} := selection \ (P_o^{(g)}, P_p^{(g)}, \mu)$ *select* from the previous and current sets

$g := g + 1$
until *termination_condition*

The mutation operation for (real) variables set y (*y_mutation*) and for strategy set, S (*s_mutation*) differs. A simple variable mutation with a constant mutation strength, σ is given as

$$\breve{y} := y + z, \text{ with } z := \sigma\left(N_1(0,1),...,N_N(0,1)\right)$$

where the $N_i (0, 1)$ are independent random samples from the standard normal distribution. However, with varying mutation strength vector,

$$z := \left(\sigma_1 N_1(0,1),...,\sigma_N N_N(0,1)\right).$$

In a more general case we have

$$z := M\left(\sigma_1 N_1(0,1),...,\sigma_N N_N(0,1)\right)$$

where M is a (orthogonal) rotation matrix. The matrix M introduces correlations between the components of z and adds $N(N - 1)/2$ parameters to the strategy set. For binary search spaces the mutation operation is carried out by flipping the bits with a probability of *mutation rate*, p_m. Strategy parameters are also mutated as below (constant mutation strength parameter):

$$\breve{\sigma}_l := \sigma_l e^{\tau N_l(0,1)}$$

Here τ, a learning parameter, is exogenous strategy parameter (to be decided by the user) and can be taken as $\tau = 1/\sqrt{N}$ or $1/\sqrt{2N}$ (for multimodal cases) [11]. The above technique can be extended to the case where we have a vector of strategy parameters

$$s = \sigma = \left(\sigma_1,...,\sigma_N\right).$$

10.3.1.3 Evolutionary programming (EP)

In EP the individuals basically comprise the variables set and the fitness is usually a scaled version of the objective function with some randomness parameter added to it. However, in the later versions of EP the individuals include a variance vector similar to that of ES. The Gaussian mutation operation is similar to that of ES except for the scaling factor, which is taken as a linear transformation of the square root of the objective function value in EP. As mutation alone is considered to create a complete new individual ready to reproduce, recombination is not done in EP. The selection is done from the combined population of the parent and the mutated population. For each individual k, q individuals are randomly chosen and scored according to how many of the q are inferior to the k. After completing this process for all the individuals in the combined pool, the top half of the individuals with highest score is selected as the population of the next generation.

10.3.1.4 Differential evolution (DE)

DE is similar to GA with some differences in coding, crossover and mutation operations [6]. An individual is represented as a series of real values representing all the decision variables of the problem. Mutation creates a new 'donor' individual (y_i) from a set of three randomly chosen individuals (x_p, x_q, x_r):

$$y_i^{t+1} = x_p^t + F\left(x_q^t - x_r^t\right)$$

where $F \in [0, 1]$ is a parameter called *differential weight*. The crossover operator, using a crossover probability C_r, generates a new individual from the portions of donor individual and existing individual. For the simple binomial crossover:

for all j : $j = 1$ to d $(d = $ number of decision variables$)$
$z_{j,i}^{t+1} = z_{j,p}^{t+1}$ if $r_i \leq C_r$
           $= x_{j,p}^t$ otherwise

where $r_i \in N(0,1)$ is a random number. In the exponential crossover scheme, a random segment, with random length, of the donor vector is selected and considered to be taken into the new individual.

10.3.2 Other nature-inspired algorithms

Apart from the evolutionary algorithms discussed above, there are a few dozen nature-inspired algorithms available in the literature [12]. However, we shall limit our discussion to three major streams: particle swarm optimization (PSO) [13], ant colony optimization (ACO) [14], and cuckoo search (CS) [15].

10.3.2.1 Particle swarm optimization

Particle swarm optimization (PSO) is a population-based stochastic search method based on the collective learning and intelligence of a swarm of particles, inspired by flocks of birds searching for food in nature. Originally developed by Kennedy and Eberhart [13], PSO has gone through several refinements [16] to avoid local optimum and improve convergence rate. The principles needed to create artificial swarm intelligence as proposed by Millonas [17] are:

1. Proximity principle: the swarm should be able to do simple space and time computations.
2. Quality principle: the swarm should be able to respond to quality factors such as quality of the food or safety of the location.
3. Principle of diverse response: the swarm should not commit its activities along excessively narrow domains to guard against environmental fluctuations.
4. Principle of stability: without a worthwhile return the swarm should not change its behavior from one mode to another following the changes in environment.
5. Principle of adaptability: the swarm should be able to switch its mode of behavior when the rewards are worth the energy of switching modes.

The PSO algorithm consists of a population of d-dimensional (d corresponds to number of decision variables) particles, each with a position and a velocity vector, constantly moving around the search space and converging to the most promising regions. Particle positions and velocities are updated towards the beneficial regions observed and randomness introduced in the process keeps the exploration robust.

PSO algorithm:

$t = 0$;
initialize $x_i^0, v_i^0, (i = 1$ *to* $n; x_i, v_i \in R^d)$ — initialize d-dim positions and velocities of population of size n

compute $f\left(x_i^0\right) \forall i$ — $f(x_i, x_i \in R^d)$ is objective function (minimization)

set $x_i^* = f\left(x_i^0\right) \forall i$ — x_i^* is best solution so far of x_i

$g^* = \min\limits_i \left(x_i^*\right)$ — best solution so far across the swarm (global)

while (! *end criterion*)
$t = t + 1$
$v_i^{t+1} = \omega v_i^t + \alpha \varepsilon_1 [x_i^* - x_i^t] + \beta \varepsilon_2 [g^* - x_i^t]$ — update velocity vector. ω (0.5~0.9) is inertial constant; α (~2), β (~2) acceleration constants; ε_1 ε_2 are random numbers (0 to 1)

$x_i^{t+1} = x_i^t + v_i^{t+1}$
compute $f\left(x_i^{t+1}\right) \forall i$

$$if\ f\left(x_i^{t+1}\right) < f\left(x_i^*\right) x_i^* := x_i^{t+1} \forall i$$

update best solution of each particle

$$g^* = \min_i\left(x_i^*\right)$$

update global best

end

10.3.2.2 Ant colony optimization (ACO)

Ant colony optimization mimics the foraging behavior of ants that deposit pheromone on paths to the food so that other ants can follow the path to the food. The implication is that the intensity of the pheromone deposition is proportional to the favorability of the path, such that ants find the shortest path from their nest to the food. Evidently, this process can be exploited for solving discrete combinatorial optimization problems, such as the traveling salesman problem (TSP), routing problems, scheduling problems, etc. As many of these are NP-hard problems – those that cannot be solved exactly in polynomial time – ACO is useful for finding good-quality solutions at a fraction of the computational cost needed by exhaustive methods [14]. ACO consists of a population of ants, each of which builds a solution (the sequence of cities to be visited in the TSP) in a step-by-step manner. The steps taken by an ant (a step in the sequence in the TSP) depends on the pheromone density related to that step, with some randomness incorporated. Once various solutions are traced by the populations of ants, the pheromone level of each step is updated based on the frequency of that particular step used by different ants, along with other metrics such as the length of the step (distance between the cities in the TSP). This completes one iteration and the process continues till a given termination condition is met.

Basic ACO algorithm:

$$\tau_{i,j} = 0 \ \forall \ (i, j) : i, j \in n, i \neq j$$

initialize pheromone levels between all the pairs of n stages $(s_1, ..s_n)$ of the problem

while (! *termination condition*)

    iter = 1

    for k = 1 to m   construct solutions for m ants

    $A_{k-1} := s_i$, randomly pick a stage $s_i \in (s_1,..,s_n)$ pick first stage with uniform probability

    for r = 2 to n

$$A_{k,r} := s_j \colon p_{ij}^k = \frac{\tau_{ij}^{\alpha} . \eta_{ij}^{\beta}}{\sum_{c_{i,l} \in \mathrm{N}(s^p)} \tau_{i,l}^{\alpha} . \eta_{i,l}^{\beta}} \ if c_{i,j} \in \mathrm{N}\left(s^p\right)$$

$\prod_{r=1}^{n} A_{k,r}$ is solution constructed by ant k

$$= 0 \ otherwise$$

from a stage 'i' a stage 'j' is selected with

$i := j$ (realized value)

probability p_{ij}^k. $N(s^p)$ denotes the set of allowed stages after passing r stages; α, β are parameters, $\eta = 1/d_{ij}$ (distance between stages i and j)

$$\tau_{i,j} := \left(1-\rho\right)\tau_{i,j} + \sum_{k=1}^{m}\Delta\tau_{ij}^k$$

update pheromone values.

$$\Delta\tau_{ij}^k = \frac{Q}{L_k} \text{ if ant } k \text{ used edge }(i,j)$$

ρ is pheromone evapora-tion rate

$$= 0 \text{ otherwise}$$

$\Delta\tau_{ij}^k$ is the pheromone laid by ant k on edge (i,j); Q is a constant; L_k is the total length of the solu-tion constructed by ant k

iter := *iter* + 1

10.3.2.3 Cuckoo search

Cuckoo search (CS) is another stochastic search mechanism consisting of both local and global look-outs for promising solutions [15]. The algorithm consists of initializing and evaluating a population of solutions and improv-ing these solutions in each iteration by means of global and local searches. Each iteration consists of a global search with the random step size based on Lévy flight distribution – a heavy-tailed distribution which throws up unex-pected or outlier values in samples – followed by replacement of an inferior solution of the population with the newly found solution. Further, a fraction of the population is replaced by better solutions found by conducting a local search near the existing solutions. In a comparative study of nature-based algorithms conducted by Wang et al. [7], CS was found to be a top performer.

Basic CS algorithm:

$t = 0$
initialize $x_i^0 (i = 1 \text{ to } n; x_i \in R^d)$

initialize 'n' individuals in d-dimensional space ('d' decision variables)

compute $f\left(x_i^0\right)\forall i$

$f(x_i, x_i \in R^d)$ is objective function (min'ze)

while (! *termination condition*)
select an i randomly
$x_*^{t+1} = x_i^t + \alpha L\left(\lambda\right)$

$L(\lambda)$ is Lévy step with exponent $\lambda \sim 1.5$; α is step size scaling factor ~ 1

$$L(\lambda) \approx \frac{U}{|V|1/\lambda}; \ U \sim N(0,\sigma^2), \ V \sim N(0,1)$$

$N(0,\sigma^2)$ is normal distri'n;, σ^2 is variance

$$\sigma^2 = \left[\frac{\Gamma(1+\lambda)}{\lambda\Gamma((1+\lambda)/2)} \cdot \frac{\sin(\pi\lambda/2)}{2^{(\lambda-1)/2}} \right]^{1/\lambda}$$

$\Gamma(\ )$ is gamma function

select a solution, say j, randomly among n

$$if \ f\left(x_*^{t+1}\right) < f\left(x_j^t\right) x_j^t := x_*^{t+1}$$

replace an inferior solution with the new one

$$Rank \ f\left(x_i^t\right) \& \ select \ \text{bottom}(p_a.n)$$

$$x_i^{t+1} = x_i^t + \alpha.s.H\left(p_a - \varepsilon\right).\left(x_j^t - x_k^t\right)$$

replace $p_a \sim 0.25$ fraction of inferior

$\forall i \in$ bottom $(p_a.n)$ (with random j,k)

solutions; ε is uniform random number in $(0,1)$; s is step size $(0.1\sim1)$

$H(+/-\ ) = 1/0$: *Heaviside/step function*

$$x_i^{t+1} = x_i^t \ \forall i \in \text{top} \ (n\text{-}p_a.n),$$

keep $(n\text{-}p_a.n)$ top solutions

$t = t + 1$

10.3.3 Artificial neural networks

Artificial neural networks (ANN) mimic the process carried out in the network of biological neuron cells. In a biological neuron, dendrites receive input from several connected neurons through synapses (junctions), and the strength of the input gets modified according to the strength of the synaptic connections (Figure 10.3). If the gross input received by the neuron crosses a certain activation threshold, the neuron 'fires', sending outputs (signals) to the connecting neurons through its axon. In response to the external stimuli the synaptic strengths and the activation threshold of the neurons get modified, such that a certain set of inputs to the network generates a certain output, resulting in 'learning' of a task by the network. Though the actual working of biological neural networks can be a lot more complex than the

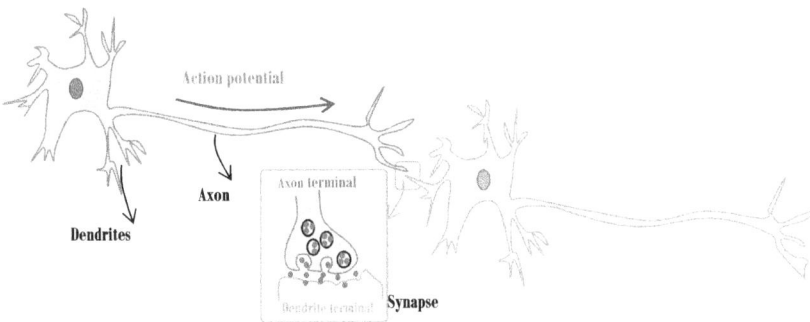

Figure 10.3 Neuron cells and synaptic junction.

Input nodes / Layer

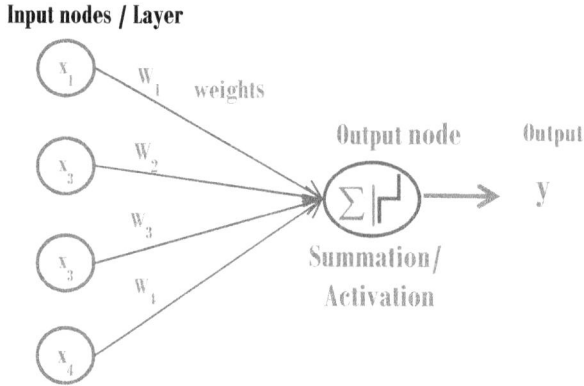

Figure 10.4 Perceptron.

above simple learning mechanism, it can be readily exploited by a network of nodes in ANN from propagating inputs to generating the required output.

In ANN a perceptron (Figure 10.4) – the simplest neural network – receives inputs from a number of nodes and sums them up after multiplying by the respective connection strengths. Based on the gross input received, the activation function of the perceptron generates the output. A more capable network will have more layers of neurons (Figure 10.5), consisting of an input layer, hidden layers and an output layer. ANN is a powerful tool for function approximation and pattern recognition and is used in certain types of optimization problems such as combinatorial optimization. ANNs can capture any complex function and in fact, representation of more complex functions can be achieved by adding more hidden layers and more nodes to the network, forming a 'deep network'.

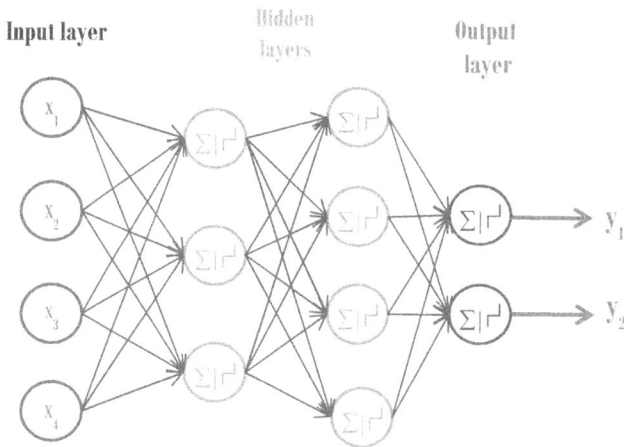

Figure 10.5 Multilayer feed-forward neural network.

Training of the network is accomplished by tuning the weights of the connections, using a given training data set. Activation functions, number of layers, nodes in each layer, etc., are parameters of the network to be decided beforehand. In supervised learning, the training set consists of pairs of input and output and the weights are adjusted so that the output obtained in the network matches the target output of the data set. The difference between the computed output and the target output, or a 'loss function' based on this difference is computed and adjustment of the weights is carried out as per the gradient of the loss function with respect to these weights. As the weights are adjusted after the loss function and gradients of it with respect to the weights are computed from output layer to input layer, this training method is called backpropagation. The trained network can be used for generating function evaluations or predicting output for the input presented. In unsupervised learning, there is no *a priori* target for the output and it is used for clustering the given data or for reducing the dimensionality of the data, etc.

The most common neural network architecture is the feed-forward network (Figure 10.5), where the input layers feed hidden layers which in turn feed output layers in a sequential manner. Radial basis function networks, Hopfield networks, restricted Boltzmann machines, recurrent neural networks, and convolutional neural networks are the other architectures, differing in the number of their layers, nature of their connections, type of training, method of learning, etc.

ANN primarily utilizes optimization schemes to minimize the loss function value to determine the best weights during training. Conversely, however, ANNs can also be used directly or indirectly in solving an optimization problem. A feed-forward ANN can be efficiently used to simplify the functions in an optimization so that the simplified functions are amenable to regular optimization algorithms. Another class of ANNs, known as Hopfield networks, can be used for a wide class of combinatorial optimization problems.

10.3.3.1 Feed-forward network for objective function simplification

ANN can be used to transform a complex objective function to a polynomial one that can be handled easily by regular optimization methods. However, we need to generate a training set for the objective function to be captured by the neural network. We employ a backpropagation algorithm for training the feed-forward network. Considering a simple three-layer feed-forward network with an input layer containing two units (Figure 10.6), one hidden layer with two units and an output layer with one unit, the procedure for backpropagation training is as follows:

$$z_j = \sigma\left(\sum_i x_i w_{ij}^1\right) j = 1, 2 \tag{10.9}$$

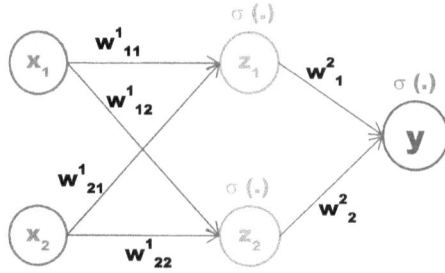

Figure 10.6 A simple feed-forward neural network.

Consider activation function as sigmoid function, σ

$$\sigma(x) = 1/\left(1 + e^{-x}\right) \tag{10.10}$$

We have

$$\frac{d\sigma(x)}{dx} = \sigma(x)\left(1 - \sigma(x)\right) \tag{10.11}$$

$$y = \sigma\left(\Sigma_k z_k w_k^2\right) \text{ OR } y = \sigma\left(\Sigma_k \sigma\left(\Sigma_i x_i w_{ik}^1\right) w_k^2\right) \tag{10.12}$$

Define the loss (or error) function as: $L = \frac{1}{2}\,(t - y)^2$ where y is the final output from the network and t is the target.

In the backpropagation method, starting with a random set of weights, the weights will be updated as follows in each iteration till the termination condition is met.

$$w_{(\text{new})} = w_{(\text{old})} - \eta.\partial L/\partial w_{(\text{old})}. \tag{10.13}$$

Weights in layer-2 (w_k^2) and layer-1 (w_{ij}^1) are updated according to chain rule. η is learning rate (− sign is to minimize the loss)

$$\frac{\partial L}{\partial y} = -(t - y) \tag{10.14}$$

$$\frac{\partial L}{\partial w_k^2} = \frac{\partial L}{\partial y} \cdot \frac{\partial y}{\partial w_k^2} = -(t - y).y(1 - y).z_k \ \forall k \tag{10.15}$$

$$\frac{\partial L}{\partial w_{ij}^1} = \frac{\partial L}{\partial y} \cdot \frac{\partial y}{\partial z_j} \cdot \frac{\partial z_j}{\partial w_{ij}^1} = -(t - y).z_j(1 - z_j).x_i \ \forall i, j \tag{10.16}$$

10.3.3.2 Hopfield networks for combinatorial optimization problems

Hopfield networks, based on the Hopfield–Tank model (H–T model), are useful for solving combinatorial optimization problems such as TSP [18]. Consider a system of N fully inter-connected neurons with internal state (sum of the weighted inputs) as u_i and external state as v_i (after applying the activation function).

$$u_i(t+1) = \sum_{j=1}^{N} W_{ij} v_j(t) + I_i \tag{10.17}$$

where I_i is constant external input (bias)

$$v_i(t+1) = f(u_i) = 1 \ if \ u_i > 0 \tag{10.18}$$

$$= 0 \ if \ u_i \leq 0$$

where $f(.)$ is activation function;
$w_{ii} = 0$ and $w_{ij} = w_{ji}$ (symmetric weights matrix)

Note that in the above representation W_{ij} weights are constant (for a different purpose involving creation of a memory state, the weights will be trained so that the network can retrieve the remembered state given a partial input vector) and only external and internal states of the neurons change dynamically with random updates. As the neurons update, (provided the weights are symmetric and non-negative diagonal elements) the 'energy' function E_d given below is guaranteed to be minimized until the system converges to a stable state (local minima):

$$E_d = -\frac{1}{2} \sum_{i=1}^{N} \sum_{j=1}^{N} W_{ij} v_i v_j + \sum_{i=1}^{N} v_i I_i \tag{10.19}$$

We can observe that the energy function E_d will be minimized if v_i are updated by Eqs. 10.17 and 10.18, which perform gradient descent on E_d. It is evident that a proper energy function needs to be built for constructing a Hopfield network to solve a combinatorial problem. Let us see the procedure for solving $N - city$ TSP using quadratic formulation [19].

Consider the binary decision variable

$X_{ij} = 1$ if city i is in position j of the tour
$= 0$ otherwise

Let d_{ik} be the distance between the cities i and k. We have

$$\text{minimize} \sum_{i=1}^{N} \sum_{k=1, k \neq i}^{N} \sum_{j=1}^{N} d_{ik} X_{ij} \left(X_{k,j+1} + X_{k,j-1} \right) \tag{10.20}$$

$$\text{subject to } \sum_{i=1}^{N} X_{ij} = 1 \, \forall \, j \tag{10.21}$$

$$\sum_{j=1}^{N} X_{ij} = 1 \, \forall \, i \tag{10.22}$$

$$X_{ij} \in (0,1) \forall \, i, j$$

We can see that in Eq. 10.20, d_{ik} gets added for a tour consisting of city i in position j to city k ($\neq i$) in position $j + 1$ or $j - 1$. Eq. 10.21 restricts only one city to be reached in a position and Eq. 10.22 restricts placing a city in one position only. We need to construct an unconstrained formulation using the objective function and the penalty terms for the constraints (we can just as well use Lagrange multipliers).

The penalty term for Eq. 10.21 can be considered as $\sum_{i=1}^{N} \sum_{k=1,k\neq i}^{N} X_{ij}X_{kj} \forall j$, which will be zero if city j is not reached from both cities i and k. Similarly, the penalty term for Eq. 10.22 can be considered as $\sum_{j=1}^{N} \sum_{l=1 l\neq j}^{N} X_{ij}X_{il} \forall i$. Now we can write the Hopfield energy function for the TSP as:

$$E = \frac{A}{2} \sum_{j=1}^{N} \sum_{i=1}^{N} \sum_{k=1,k\neq i}^{N} X_{ij}X_{kj} + \frac{B}{2} \times \sum_{i=1}^{N} \sum_{j=1}^{N} \sum_{l=1,l\neq j}^{N} X_{ij}X_{il}$$

$$+ \frac{C}{2} \left(\sum_{i=1}^{N} \sum_{j=1}^{N} X_{ij} - N \right)^2 + \frac{D}{2} \times \sum_{i=1}^{N} \sum_{k=1,k\neq i}^{N} \sum_{j=1}^{N} d_{ik}X_{ij} \times \left(X_{k,j+1} + X_{k,j-1} \right) \tag{10.23}$$

The third term in the above forces each city to occupy one position necessarily. The penalty parameters A, B, C and D need to be balanced to meet the constraints and, at the same time, ensure that the solution is the optimum. Next, we need to obtain the network weights and biases so that the energy function is minimized by the update rules. Eq. 10.23 needs to be written in the format of Eq. 10.19 as follows:

$$E_d = -\frac{1}{2} \sum_{i=1}^{N} \sum_{j=1}^{N} \sum_{k=1}^{N} \sum_{l=1}^{N} W_{ijkl}X_{ij}X_{kl} - \sum_{i=1}^{N} \sum_{j=1}^{N} I_{ij}X_{ij} \tag{10.24}$$

W_{ijkl} in Eq. 10.24 compares with W_{ij} in Eq. 10.19, except that in Eq. 10.19 the neurons can be said to be one-dimensional, whereas in Eq. 10.24 they are two-dimensional, since each neuron is identified with the index of the city and the position (Figure 10.7). To find out the weights, etc. from Eq. 10.24, we need to expand Eq. 10.23, matching the format of Eq. 10.24.

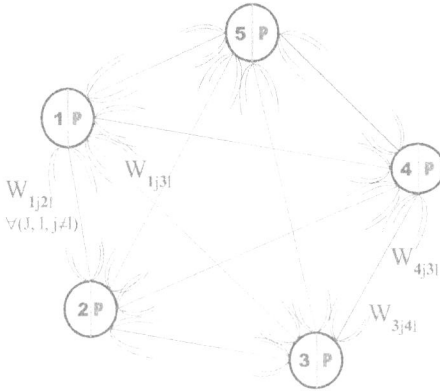

Figure 10.7 Hopfield network for five-city TSP problem. Each neuron's external state (X_{ij}) is a two-dimensional state; one dimension (i) holds city index and the other (j) holds city position. Neurons with city indices are shown above, where **P** shows all the possible positions that can be taken up by the neuron, which is five here. W_{1j2l} indicates the weight connecting from city 1 taking a position of j to city 2 taking a position of l (j, l: 1 to 5, $l \neq j$). So we will have 20 connections from city 1 to city 2 (weights are symmetric or $W_{1j2l} = W_{2j1l}$). At the minimum energy level, the solution is **(i,j): $X_{ij} = 1$**.

$$E = -\frac{1}{2}\sum_{i=1}^{N}\sum_{j=1}^{N}\sum_{k=1}^{N}\sum_{l=1}^{N}[-A\delta_{jl}\left(1-\delta_{ik}\right)-B\delta_{ik}\left(1-\delta_{jl}\right)-C-Dd_{ik}\left(1-\delta_{ik}\right)$$

$$\times\left(\delta_{l,j+1}+\delta_{l,j-1}\right)]X_{ij}X_{kl}-\sum_{i=1}^{N}\sum_{j=1}^{N}[CN]X_{ij}+\frac{CN^2}{2} \qquad (10.25)$$

where δ_{ij} is Kronecker-Delta: $\delta_{ij} = 1$ if $i = j$ and $\delta_{ij} = 0$ if $i \neq j$.

Comparing Eqs. 10.24 and 10.25, and ignoring the constant term in Eq. 10.25 which doesn't affect the local minima, we can write the network parameters as:

$$W_{ijkl} = -A\delta_{jl}\left(1-\delta_{ik}\right)-B\delta_{ik}\left(1-\delta_{jl}\right)-C-Dd_{ik}\left(1-\delta_{ik}\right)\left(\delta_{l,j+1}+\delta_{l,j-1}\right)$$

$$I_{ij} = CN \qquad (10.26)$$

With the above parameters the Hopfield network can be initialized with random values for the initial states and updated according to the update rules of Eqs. 10.17 and 10.18, replacing v_i with X_{ij}. The network is guaranteed to reach a stable local minimum of the energy function which optimizes the objective function of the TSP.

Hopfield networks can be used for other OR&O problems such as scheduling, sorting and networking. A continuous version of the H–T model can solve linear programming and nonlinear programming problems for local minimum though this is not a popular or best option. As can be noted here, the Hopfield network needs a suitable energy function with several parameters to be constructed and performs gradient descent to reach only local minima. Many successful studies improving the construction of energy function and selection of parameters are reported in the literature [20]. It has also been found that incorporation of metaheuristic methods into the H–T model can achieve global convergence. Moreover, ANN, apart from serving as a computational algorithm in a digital computer, can be implemented in real-time analog circuits.

10.4 CONCLUSION

Various types of OR problems and the resulting optimization models were discussed in the context of obtaining optimal solutions within the available computational resources. AI-based techniques such as evolutionary algorithms, nature-based algorithms and artificial neural networks (ANNs) are presented as practical and efficient method of obtaining optimal or near-optimal solutions to computationally heavy OR&O problems.

REFERENCES

1. Rechenberg I. *Cybernetic solution path of an experimental problem. Roy. Aircr. Establ., libr transl. 1122.* Hants, U.K.: Farnborough; 1965.
2. Rechenberg I. *Evolutionsstrategie: Optimierung technischer Systeme nach Prinzipien der biologischen Evolution.* Stuttgart: Frommann-Holzboog; 1973.
3. Goldberg DE. *Genetic algorithms in search, optimisation and machine learning.* Reading, MA, USA: Addison-Wesley; 1989.
4. Holland J. *Adaptation in natural and artificial systems.* Ann Arbor, MI, USA: University of Michigan Press; 1975.
5. Fogel LJ, Owens AJ, and Walsh MJ. *Artificial intelligence through simulated evolution.* New York: John Wiley; 1966.
6. Storn R and Price K. Differential evolution: a simple and efficient heuristic for global optimization over continuous spaces. *J Glob Optim*; 11:341–59; 1997.
7. Wang Z, Qin C, Wan B, and Song WW. A Comparative Study of Common Nature-Inspired Algorithms for Continuous Function Optimization. *Entropy*; 23, 874; 2021.
8. Michalewicz Z. *Genetic algorithms + data structures = evolution programs.* Berlin: Springer; 1992.
9. Srinivas N and Deb K. Multiple objective optimization using nondominated sorting in genetic algorithms. *Evol Comput*; 2(3):221–48; 1995.
10. Deb K, Agrawal S, Pratap A, and Meyarivan T. A fast and elitist multi-objective genetic algorithm: NSGA-II. *IEEE Trans Evol Comput*; 6(2):182–97; 2002.

11. Beyer H-G and Schwefel H-P. Evolution strategies – A comprehensive introduction. *Nat Comput*; 1:3–52; 2002.
12. Yang XS. *Nature-inspired optimization algorithms*. First ed. Elsevier; 2014.
13. Kennedy J and Eberhart RC. Particle swarm optimization. In: *Proceedings of the IEEE International Conference on Neural Networks*, Piscataway, NJ, USA; pp. 1942–8; 1995.
14. Dorigo M, Birattari M, and Stützle T. Ant colony optimization. *IEEE Comput Intell Mag*; 1(4):28–39, November 2006.
15. Yang XS and Deb S. Cuckoo search via Lévy flights. In: *Proceedings of world congress on nature & biologically inspired computing* (*NaBIC 2009*). USA: IEEE Publications; pp. 210–4; 2009.
16. Wang D, Tan D, and Liu L. Particle swarm optimization algorithm: an overview. *Soft Comput*; 22:387–408; 2018, Springer; published online: 17-Jan-2017.
17. Millonas MM. Swarms, phase transitions, and collective intelligence. SFI Working Paper: 1993-06-039 (paper 1); 1993.
18. Hopfield JJ and Tank DW. Neural computation of decisions in optimization problems. *Biol Cybern*; 52(3):141–52; 1985.
19. Potvin JY and Smith KA. Artificial Neural Networks for Combinatorial Optimization. In: Glover F and Kochenberger, GA. (eds) *Handbook of Metaheuristics*. International Series in Operations Research & Management Science, vol. 57. pp. 429–55, Springer, Boston, MA; 2003.
20. Syed MN and Pardalos PM. Neural Network Models in Combinatorial Optimization. In Pardalos PM et al. (eds) *Handbook of Combinatorial Optimization. 2027-2093.* Springer Science + Business Media, New York; 2013.

Chapter 11

Heuristic strategies for warehouse location with store incompatibilities in supply chains

Adarsh Pandey
Cadence Design Systems (India) Pvt. Ltd., Noida, India

Nidhip Taneja
Keysight Technologies, Gurugram, India

Prem Prakash Vuppuluri
Dayalbagh Educational Institute (Deemed University), Agra, India

CONTENTS

11.1 INTRODUCTION

Facility location problems (FLPs) constitute a class of well-known classic optimization problems in which the objective is to determine the best locations for establishing a set of facilities such as factories or warehouses based

DOI: 10.1201/9781003415466-11

on geographical demands, facility costs and transportation distances, while maximizing some parameter such as the supplier's profits, given a set of user/customer demands, locations and associated costs. Effective decision-making in terms of the number and location of facilities plays a significant role in improving capital and inventory costs, which makes FLPs crucial for effective strategic planning, design and deployment of systems in organizations all over the world today. FLPs are viewed as being either capacitated or uncapacitated: capacitated facility location problems have constraints in terms of the quantity of goods that can be produced and transported by each facility, whereas uncapacitated FLPs do not. Hence, in an uncapacitated FLP, in the absence of any further constraints, the demand for goods would always be supplied from the closest/lowest-cost facility. In a capacitated FLP, it is possible that a facility might not contain sufficient goods to satisfy the nearest demand.

FLPs are commonly formulated as mixed-integer linear programs (MILPs) having a set of locations for facilities and customers. The availability (or lack thereof) of facilities and their ability to supply items to each customer are typically represented in such programs as binary variables. In a problem with N warehouses and M stores, the capacitated formulation defines a binary variable x_i and a variable y_{ij} for each warehouse i and each store j. If warehouse i is open, $x_i = 1$; otherwise $x_i = 0$. Open warehouses have an associated fixed cost f_i and a maximum capacity $k_i.y_{ij}$ is the fraction of the total demand d_j of store j that warehouse i has satisfied and the transportation cost between warehouse i and store j is represented as t_{ij}.

The capacitated FLP is therefore defined as

$$\min \sum_{i=1}^{N}\sum_{j=1}^{M} d_j t_{ij} y_{ij} + \sum_{i=1}^{N} f_i x_i$$

$$\text{s.t.} \sum_{i=1}^{N} y_{ij} = 1 \, \forall j \in \{1...M\}$$

$$\sum_{j=1}^{M} d_j y_{ij} \leq k_i x_i \, \forall i \in \{1...N\}$$

$$y_{ij} \geq 0 \, \forall i \in \{1...N\}, \forall j \in \{1...M\}$$

$$x_i \in \{0,1\} \, \forall i \in \{1...N\}$$

FLPs and their variants have been extensively studied over the years and find wide application in several real-life problem domains such as military supply chains and logistics [1], healthcare [2], disaster management [3], distributed generation placement in power systems [4], drone-based goods delivery [5], and so on.

Most variants of FLPs are known to be NP-hard, implying that exact algorithms for these problems would require computational time that grows exponentially with problem size, which in turn encourages the development of heuristic and meta-heuristic strategies that are able to obtain good solutions within acceptable time limits and scale well for practical problem sizes. One of the earliest solutions to the (capacitated) FLP (CFLP) is a branch-and-bound algorithm given by Umit Akinc and Basheer M. Khumawala [6] in 1977. A variant of the CFLP called generalized FLP (GFLP) is presented in [7], where the setup costs comprise a combination of the costs incurred in setting up a site and (multiple) facilities in the site. The costs are non-linear in terms of the site size. Two mixed-integer linear programming (MILP) models and a Lagrangian heuristic are presented for the GFLP, and their performance evaluated several randomly generated test instances as well as existing benchmarks from Beasley's OR library [8]. A revised Vogel-based algorithm called dynamic VAM (DVAM) is proposed for a multi-level capacitated facility location problem (MCFLP) [9] that extends the well-studied two-level facility location problem [10]. In this problem, goods are routed from source to destination via multiple intermediate facilities, and the goal is to open facilities across the distribution levels such that client demands are satisfied and the overall distribution cost is minimal. The MCFLP problem finds application in areas such as telecommunications and hierarchical logistics. DVAM is shown to outperform other heuristics such as LINGO (based branch-and-bound algorithm) and the 'greedy method'. A real-life application of the facility location problem for solid waste management in Lagos, Nigeria is studied in [11]. The focus in this work is to find effective ways to situate waste removal centers so as to minimize facility setup costs, total distance and overall transportation costs incurred. A solution to the problem based on the particle swarm optimization algorithm is presented in [12]. Several problem-specific heuristics have been developed for FLP variants over the years, a few of which have been briefly discussed in this section. For a more in-depth review of models and algorithmic techniques for various FLP variants and their applications, the interested reader is referred to comprehensive surveys of recent research developments such as [13–15] or [16].

In this chapter, we study the effectiveness of using two different heuristic strategies for solving a recently proposed FLP variant known as the 'Capacitated facility location problem with store incompatibilities' [17]. The first of the two proposed heuristics modifies the well-known Vogel's Approximation Method [18] for the problem at hand, while the second heuristic builds solutions using 'greedy' strategies for store selection and warehouse allocation. The remaining sections of the chapter are organized as follows: the warehouse location problem with store incompatibilities is introduced in Section 11.2, and the proposed heuristics are described in Sections 11.3 and 11.4. Experimental work is presented in Section 11.5 and conclusions are drawn in Section 11.6.

11.2 CAPACITATED FACILITY LOCATION WITH STORE INCOMPATIBILITIES

Consider a set W of warehouses (or more generally, facilities), each having a finite *capacity* that represents the maximum number of goods that it can hold (we assume that all goods are of the same type), and a fixed, one-time *setup cost* or *opening cost* incurred for opening the warehouse. Opened warehouses supply goods to a set S of stores, each of which has a *demand* for a certain quantity or measure of goods that may be satisfied by one or more warehouses. There is also a cost associated with the supply of goods from a warehouse w to a store i; this is simply the product of the number of items supplied with the unit cost for supplying one item from w to i. The supply cost varies depending on the specific warehouse and store in question. For instance, supply costs to a store located near the warehouse could be significantly lower than to a store that is distant. In the classical warehouse location problem (WLP), the goal is to select a subset of available warehouses for opening and supplying goods to a set of stores such that the total of supply costs and warehouse opening costs incurred is minimal. The Capacitated facility location problem with store incompatibilities extends this problem by applying an additional constraint that requires that specific pairs of conflicting or *incompatible* stores may not be supplied from the same warehouse. For example, consider two stores, say i and j, that are said to conflict with one another. This means that, for a warehouse w, if goods to store i is being supplied by w, then store j may not be supplied by w, even if w could have satisfied the demand of store j.

A solution to the problem may be represented by a list of <s, w, q>, where each triple in the list represents the quantity q of goods allocated to store s from opened warehouse w. The solution must ensure that the total number of goods/items allotted to a store exactly matches that store's demand, and that this holds true for all the stores in S. Further, no warehouse may supply goods beyond its stated capacity, and goods may be supplied only from opened warehouses. Lastly, no two stores in conflict with each other receive supplies from the same warehouse. The objective of the problem is to identify the solution that minimizes the total cost incurred in opening a subset of warehouses and allocating goods from open warehouses to all the stores as per their respective demands while respecting all these constraints.

11.3 HEURISTICS BASED ON VOGEL'S APPROXIMATION METHOD (VAM)

In this section, we present three heuristics based on Vogel's Approximation Method (VAM) for solving the problem. The heuristics differ from each other in terms of how incompatible pairs of stores are selected and how the warehouses are allocated for the selected store.

Step 1: Balancing an unbalanced transportation problem. unbalanced problems occur in two different ways, when there is: (i) Excess of availability of goods, or (ii) Shortage of availability of goods. So, we try to balance by making availability and demand equal. This happens as -

    Case I - Excess of availability of goods: In this case we add one extra dummy column (of dummy stores) with each of its row initialized with 0 and demand of that dummy column initialized to (total supply – total demand).

    Case II – Shortage of availability of goods: In this case we add one extra dummy row (of dummy warehouse) with each of its column initialized with 0 and capacity of that dummy row initialized to (total demand – total supply).

Step 2: Determine the penalty cost for each row and column by subtracting the lowest cell cost in the row from the next lowest cell cost in the same row and   similarly subtracting the lowest cell cost in the column from the next lowest cell cost in the same column. Repeat for each row and each column.

Step 3: Select the row or column with the highest penalty cost.

Step 4: Select the lowest supply cost cell in the selected highest penalty row or column done in step 3.

Step 5: Allocate as much as possible (lowest among supply or demand) to the cell with the lowest supply cost.

Step 6: Repeat steps 2, 3, 4 and 5 until (total demand >0 and total supply>0).

Step 7: Compute total cost.

Figure 11.1 Pseudo-code for generalized Vogel Approximation Method.

Determining efficient solutions for large-scale transportation problems is an important task in operations research. VAM is a well-known transportation method frequently used in transportation problems for efficiently constructing good starting solutions, which can then be improved further using various other techniques. VAM usually provides starting solutions that are closer to optimum than other methods for transportation problems, in computation time that is polynomial in the input size. However, there is no guarantee that the algorithm will be able to return an optimal solution. The main steps of the general VAM algorithm are listed in Figure 11.1.

The proposed VAM-based heuristics first allocate the demands of the incompatible stores and then fulfill the demands of the remaining (non-conflicting) stores. Each heuristic uses a different approach for allocating the demands of the incompatible stores, and the same VAM-based strategy for allocating goods to the remaining stores. The proposed VAM-based heuristics are described in greater detail in the next sub-section.

11.3.1 Modified VAM heuristic (MVH1) for incompatible store pairs

The heuristic first allocates supplies to only the incompatible stores. In particular, the column penalty is computed for each store in an incompatible pair, and the store with the higher penalty is identified. Thereafter, the minimum supply cost in the selected highest penalty column is chosen and the store is allocated as many units of goods as needed. Subsequently, the general VAM approach is used for supplying goods to the remaining stores. Pseudo-code for this heuristic is given in Figure 11.2. Once the required warehouses are opened and allocation of goods from warehouses to stores is made, the total cost incurred may be computed by summing the opening costs of the warehouses that were opened and the total supply costs of allocating all the stores' demands to open warehouses. Pseudo-code for this procedure is given in Figure 11.3. For m warehouses and n stores $(n > m)$, an upper bound for the running time of the heuristic is given in Lemma 1.

```
1  dp[warehouseNo][storeNo+1]      ▷to perform all operations
2  dp2[warehouseNo][storeNo+1]     ▷to calculate final total amount by multiplying with
                                      allocations
3  allotedGoods[warehouseNo][storeNo+1]  ▷to store allocated goods
4  Goods[storeNo + 1] ← (totalCapacity − totalDemand)   ▷this we add to balance
                                                   totalDemand and totalCapacity
5  for incompatiblePairs = 1, 2, . . . , N do
6      Get column penalty for both stores in incompatible pair by subtracting the lowest cell
       from the next lowest cell in the column.
7      Select the higher penalty among two and save storeNo corresponding to that store.
8      Select min(supplyCost) in the highest penalty column and save the minimum
       supplyCost row into mincost_row.
9      Select min(Goods.capacity) corresponding to the column (column here is the selected
       store among the incompatible pair) and row is the mincost_row).
10     if Goods[storeNo] ≤ Capacity[mincost_row] then
11         allottedGoods[storeNo][mincost_row] ← Goods[storeNo]
12         Capacity[mincost_row] ← Capacity[mincost_row]−Goods[storeNo]
13         Goods[storeNo] ← 0
14         Now, make that column complete by allocating -1 to the column.
15     else
16         allottedGoods[storeNo][mincost_row] ← Capacity[mincost_row]
17         Goods[storeNo] ← Goods[storeNo] − Capacity[mincost_row]
18         Capacity[mincost_row] ← 0
19         Now, make that row complete by allocating -1 to the row.
20     end if
21     j ← i
22     for incompatiblePairs = 1, 2, . . . , N do
23         if IncompatiblePairs[j][0] == selectedStoreNo then
24             dp[mincost_row][IncompatiblePairs[j][1]] ← −1
25         else if IncompatiblePairs[j][1] == selectedStoreNo then
26             dp[mincost_row][IncompatiblePairs[j][0]] ← −1
27         end if
28         j ← j+1
29     end for
30     i ← i+1
31 end for
32 Apply general VAM heuristic on remaining stores      ▷ common routine for all variants
```

Figure 11.2 Pseudo-code for the modified VAM heuristic for incompatible pairs.

```
 1  OpenWarehouseArr[warehouseNo] ← 0          ▷For opening the warehouse
 2  for i = 0, 1, . . . .(warehouseNo − 1) do
 3      for j = 0, 1, . . . .(storeNo − 1) do
 4          if allotedGoods[i][j] ! =0 then
 5              OpenWarehouseArr[i] ← 1
 6                  break
 7          end if
 8      end for
 9  end for
10  supply_cost ← 0     ▷Calculating total cost without including fixed cost
11  for i = 0, 1, . . . .(warehouseNo − 1) do
12      for j = 0, 1, . . . .(storeNo) do
13              supply_cost ← supply_cost + dp2[i][j] · allotedGoods[i][j]
14      end for
15  end for
16  totalCost ← supply_cost     ▷Calculating total cost with including fixed cost
17  for i = 0, 1, . . . .(warehouseNo − 1) do
18      if OpenWarehouseArr[i] == 1 then
19              totalCost ← totalCost + FixedCost[i]
20      end if
21  end for
```

Figure 11.3 Pseudo-code to total cost computation (step 7 of VAM-based heuristic).

Lemma 1

The running time of the modified VAM-based heuristic for incompatible store pairs is bounded by $O(n^3)$, where n is the number of stores and m the number of warehouses, $n > m$.

Proof. The time taken for computing the penalty associated with incompatible store pairs is $O(m)$, and the time for finding the minimum supply cost cell is also $O(m)$. Marking a row or column as complete takes $O(n)$, and the time required for flagging all incompatible stores with respect to a given store s (in order that those stores may not be considered for further allocations from the warehouse that supplies goods to store s) is $O(I)$. These operations are repeated for each of the incompatible pairs; assuming I such pairs, the total time taken for allocating the incompatible pairs works out to $O[I^*\{(2m+n)+I\}]$. For the remaining (non-conflicting) stores, the time taken to calculate the total quantity of goods required by stores and the residual capacity in the warehouse is $O(n+m)$. Computing the column penalties for each column and row penalties for each row takes the same time: $O(n.m)$. Finding the highest column and row penalty requires $O(n)$ and $O(m)$ respectively, and identifying the lowest supply-cost cell in the row or column takes $O(Max\{n, m\})$, or $O(n)$ considering $n>m$. Iterating over these steps at most $O(n)$ times until all the store demands are allocated takes

$O[n^*\{(n+m)+2^*(n^*m)+n+m+n\}] \approx O\{n^*(n^*m)\}$ time. The total running time is therefore bounded by $O\{n^*(n^*m)\} + O[I^*\{(2m+n) + I\}] \approx O(n^3)$, assuming $n > m$.

11.3.2 Modified VAM heuristic (MVH2) using minimum product for incompatible pairs

The modified VAM heuristic using minimum product uses a different approach to allocating incompatible store pairs, in conjunction with the basic VAM-based approach for the remaining stores (Figure 11.1). For each incompatible pair (s_i, s_j) of stores, the product of the respective store demand and per-unit supply cost with respect to each warehouse is computed. Thereafter, the store with the minimum value of the computed product is picked and allocated as many units of the goods as possible from the corresponding warehouse. If store s_i is picked for allocation of goods, then the corresponding warehouse is marked as inaccessible to store s_j. Once all the incompatible stores are allocated in this manner, the general VAM approach is used for allocating supplies to the remaining stores as before. Using a similar reasoning as given in Lemma 1, it may be seen that the running time of the heuristic is $O(n^3)$, where n is the number of stores. Pseudo-code for this variant of the heuristic is presented in Figure 11.4.

11.3.3 Modified VAM heuristic (MVH3) using minimum supply costs

The modified VAM heuristic using minimum supply costs (MVH3) differs slightly from the MVH2 heuristic. In this heuristic, for each incompatible pair, the store whose demand may be satisfied via a lower per-unit supply cost (from some warehouse) is selected and allocated as many units as possible at that supply cost. Once all the incompatible pairs are allocated goods in accordance with their demand, the general VAM is applied for allocating goods to the remaining stores. The running time of the heuristic is $O(n^3)$.

11.4 GREEDY HEURISTICS

This section presents two deterministic greedy heuristics and two randomized greedy variants of these heuristics. The deterministic heuristics differ in the invariant used for store selection and warehouse(s) assignment to each selected store. The randomized greedy variants select stores in random order but use the corresponding greedy invariants for choosing which warehouse(s) to allocate goods to the stores from.

```
 1: dp[whnum][storeNo+1]      ▷to perform all operations
    dp2[whnum][storeNo+1]     ▷to calculate final total amount by multiplying with allocations
 2: IncPairs[1][2]            ▷List of incompatible stores
 3: allotedGoods[whnum][storeNo+1]              ▷to store allocated goods
 4: Goods[storeNo − 1] ← (totalCapacity − totalDemand)    ▷ this we add to balance
                                                    totalDemand and totalCapacity
 5: Inc_stores_prdct[whnum][2]
 6: for incompatiblePairs = 1, 2, . . . , N do
 7:     for j = 0, 1, . . . , (whnum − 1) do
 8:         if dp[j][IncPairs[i][0]] ! = -1 && Goods[IncPairs[i][0]] !=0 then
 9:              Inc_stores_prdct[j][0] ← dp[j][IncPairs[i][0]] * Goods[IncPairs[i][0]]
10:         else
11:              Inc_stores_prdct[j][0] ← MAX_VALUE
12:         end if
13:         if dp[j][IncPairs[i][1]] ! = -1 && Goods[IncPairs[i][1]] ! = 0 then
14:              Inc_stores_prdct[j][1] ← dp[j][IncPairs[i][1]] * Goods[IncPairs[i][1]]
15:         else
16:              Inc_stores_prdct[j][1] ← MAX_VALUE
17:         end if
18:     end for
19:     Now select the minimum product among Inc_stores_prdct.
20:     Select supplyCost of selected minimum product above
21:     Select min(Goods.capacity) corresponding to the column
              ▷ column = selected store from the incompatible pair; row = mincost_row
22:     if Goods[storeNo] ≤ Capacity[mincost_row] then
23:         allottedGoods[storeNo][mincost_row] ← Goods[storeNo]
24:         Capacity[mincost_row] ← Capacity[mincost_row] − Goods[storeNo]
25:         Goods[storeNo] ← 0
26:         Now, make that column complete by allocating -1 to the column.
27:     else
28:         allottedGoods[storeNo][mincost_row] ← Capacity[mincost_row]
29:         Goods[storeNo] ← Goods[storeNo] − Capacity[mincost_row]
30:         Capacity[mincost_row] ← 0
31:         Now, make that row complete by allocating -1 to the row.
32:     end if
33:     j ← 1
34:     for incompatiblePairs = 1, 2, . . . , N do
35:         if IncPairs[j][0] == selectedStoreNo then
36:              dp[mincost_row][IncPairs[j][1]] ← −1
37:         else if IncPairs[j][1] == selectedStoreNo then
38:              dp[mincost_row][IncPairs[j][0]] ← −1
39:         end if
40:         j ← j+1
41:     end for
42:     i ← i-1
43: end for
44: Apply general VAM heuristic on remaining stores       ▷ common routine for all variants
```

Figure 11.4 Pseudo-code for the modified VAM heuristic using minimum product for incompatible pairs.

Stores	1	2	3	4	5	6	7	8	9	10
Demand	12	17	5	13	20	20	17	19	11	20

Warehouses

	1	2	3	4
1	27	66	44	55
2	53	89	68	46
3	17	40	18	61
4	20	68	44	78
5	42	89	65	78
6	57	55	49	31
7	89	101	90	16
8	37	31	23	55
9	76	60	63	44
10	82	107	91	31

Stores (row label)

Figure 11.5 Store demand and supply costs' tables for the example.

11.4.1 Greedy Heuristic based on Supply Costs and Demand Product (GDH1)

In the greedy heuristic based on supply costs and demand product (GDH1), the product p_s of the per-unit supply cost (from each warehouse) and the demand for goods is computed and saved for each store i. For each iteration, the store having the minimum value of p_s out of all currently unallocated stores is selected. Once a store is selected, it is allocated the warehouse which can satisfy the store demand at the lowest supply cost while simultaneously checking for the incompatibility constraint. This is explained further using a toy instance adapted from [17]. Assume that there are four warehouses, each with some associated capacity for storing goods, and ten stores with demands for goods to be serviced. The store-wise demand for goods and per-unit supply cost to each store from different warehouses are depicted in Figure 11.5.

The corresponding 2-D matrix of (supply cost × demand) products for each store and warehouse is shown in Figure 11.6. In the matrix, rows represent stores and columns represent warehouses. In the figure, store 3 (3rd row) has the minimum product, so it is selected first and warehouse 1 is allocated to supply goods to this store. Store 4 is selected next as it has the second-lowest product (4th row, 1st column) and so on.

For m warehouses and n stores, the total running time of the heuristic works out to $O(n^2*m)$, or $O(n^3)$, assuming $n > m$. Pseudo-code for the heuristic is given in Figure 11.8.

```
[324  792  528  660
901  1513  1156  782
85  200  90  305
260  884  572  1014
840  1780  1300  1560
1140  1100  980  620
1513  1717  1530  272
703  589  437  1045
836  660  693  484
1640  2140  1820  620]
```

Figure 11.6 2-D matrix containing product of supply cost and demand.

```
[2.25 5.5 3.6666666666666665 4.583333333333333
3.1176470588235294 5.235294117647059 4.0 2.7058823529411766
3.4 8.0 3.6 12.2
1.5384615384615385 5.230769230769231 3.3846153846153846 6.0
2.1 4.45 3.25 3.9
2.85 2.75 2.45 1.55
5.235294117647059 5.9411764705882355 5.294117647058823 0.9411764705882353
1.9473684210526316 1.631578947368421 1.2105263157894737 2.8947368421052633
6.909090909090909 5.454545454545454 5.7272727272727275 4.0
4.1 5.35 4.55 1.55]
```

Figure 11.7 2-D matrix containing ratio of supply cost to demand.

11.4.2 Greedy heuristic based on supply costs-to-demand ratio (GDH2)

This is a variant of GDH1 where, instead of the product of supply cost (SC) and store demand (SD), the ratio of supply cost to store demand is considered. As before, stores are selected onebyone on the basis of minimum ratio and similarly, warehouses are allocated on the basis of the minimum ratio for each store while ensuring that incompatibilities are not violated.

Using the example in Figure 11.5, the supply-costs-to-store-demand ratio is obtained as shown in the 2-D matrix in Figure 11.7. Here, the first store selected is store 7 and warehouse 4 is allocated to it. Store 8 is selected next as it has the next smallest ratio. Stores are selected and allocated goods from warehouses sequentially in this manner.

11.4.3 Randomized greedy heuristics

Two simple randomized variants of the greedy heuristics are also implemented in this work. The first of these, named RGD1 (short for randomized variant of the GDH1 heuristic) selects stores in random order and uses the SC–SD product for warehouse allocation as in GDH1. The second, called RGD2, uses the SC–SD ratio as in the GDH2 heuristic. Since different

```
1. initialize SolutionMatrix[i][j] ← -1            ▷ initialize final solution matrix
2. for i = 0, 1, . . . , (storeNo – 1) do
3.    for j = 0, 1, . . . ,(warehouseNo – 1) do
4.       SupplyCost_Goods[i][j] =SupplyCost[i][j]*Goods[i] ▷ 2D matrix to store product
5.    end for
6. end for
7. initialize array Minimum[]                      ▷ to store minimum product store wise
8. for each store i do
9.    Minimum[i] = min(SupplyCost_Goods[i][j]), ∀ j ∈ {0, 1,...., (warehouseNo - 1)}
10. sort(Minimum)
11. for each store do
12.    v = 0
13.    while(v<No. Of Stores)
14.       MinRatio = Minimum[v];
15.       PstoreNo = Minimum.indexOf(MinRatio); ▷ store having the next min product
16.       PwarehouseNo = SupplyCost_Goods[storeNo][indexOf(MinRatio)]
          ▷ PwarehouseNo stores the warehouse having lowest SupplyCost*Goods for this store
17.       for i = 0, 1 . . .(InCompatibilities – 1) do
18.          store all the incompatible stores in list w[]
19.       end for
20.       ▷ check if the warehouse has already allocated goods to an incompatible store
21.       if (warehouse is allocated to an incompatible store) then
22.          PwarehouseNo = newloc() ▷ warehouse with next lowest product, sans violations
23.       end if
24.       while (StoreDemandFulfilled[storeNo]== FALSE)
25.          if(Goods[PstoreNo]<=Capacity[PwarehouseNo]) then
26.             SolutionMatrix[PstoreNo][PwarehouseNo]==Goods[PstoreNo];
27.             WarehouseOpen[PwarehouseNo]== TRUE;
28.             StoreDemandFullfilled[PstoreNo]== TRUE;
29.             Capacity[PwarehouseNo] = Capacity[PwarehouseNo]-Goods[PstoreNo];
30.             Goods[PstoreNo] =0;
31.          else if (Goods[PstoreNo]>Capacity[PwarehouseNo]) then
32.             SolutionMatrix[PstoreNo][PwarehouseNo]=Capacity[PwarehouseNo];
33.             Goods[PstoreNo]=Goods[PstoreNo]-Capacity[PwarehouseNo];
34.             Capacity[PwarehouseNo]=0;
35.             Find the next warehouse having next minimum product s.t. incompatibility
             constraints are not violated, store it in "PwarehouseNo"
36.             Continue until store demand is fulfilled, then set StoreDemandFulfilled[store]=TRUE
37.          end if
38.       end while
39.       v ← v + 1
40.    end while
41. end for
42. Calculate total_variable_cost by multiplying SolutionMatrix with supplycost matrix
43. for i = 0,...warehouseNo – 1 do
44.    if open(warehouse_i) then
45.       total_fixed_cost ← total_fixed_cost + opening_cost(warehouse_i)
46. Return total variable cost + total fixed cost
```

Figure 11.8 Pseudo-code for the greedy heuristic GDH1.

random sequences of stores can impact the overall cost, each heuristic is run a small constant K number of times (so as to not significantly increase the total computation time) and the best result obtained is returned.

11.5 EXPERIMENTAL WORK

A set of 30 test instances [17] is provided for the problem in standard MiniZinc data file format. The first 15 of these test instances were used for studying the performance of the proposed heuristics. Denoted as triples comprising the number of warehouses, the number of stores and the number of conflicting pairs of stores, respectively, these test instances range in size from (50, 115, 383) to (1400, 3445, 320,634). Each test instance contains the following information: the number of warehouses, the capacity and fixed opening cost of each warehouse, the number of stores, the demand for goods from each store, the per-unit cost of supplying goods to each store from each warehouse, the number of conflicting stores and the actual pairs of conflicting stores. All the heuristics described in this work were implemented in C++ on a computing system with Intel i5 (7th Gen) processor, 4 GB RAM running Windows 10 (Home Edition).

Table 11.1 presents the results obtained by the Vogel-based heuristics. For each of the 15 test instances, numbered wlp01 through wlp15, the total cost and the number of constraint violations obtained by each of the three Vogel-based heuristic variants are reported. The results for VAM1 show that the heuristic performs slightly worse than the other two on all

Table 11.1 Results obtained by the heuristics based on Vogel's Approximation Method (VAM1, VAM2 and VAM3) on the benchmark problems

Instances	VAM1	Violations	VAM2	Violations	VAM3	Violations
wlp01	43024	0	43339	0	**42674**	0
wlp02	83122	0	**80272**	0	80357	0
wlp03	110530	0	**109778**	0	109780	0
wlp04	147186	0	140417	0	**140401**	0
wlp05	173485	0	172559	0	**172547**	0
wlp06	193746	0	189409	0	**189405**	0
wlp07	267981	1	268970	0	**268405**	0
wlp08	325761	0	320838	0	**320455**	0
wlp09	392890	1	**389336**	0	389336	0
wlp11	506060	0	502251	0	**502247**	0
wlp12	554255	0	**550600**	0	550610	0
wlp13	**585397**	0	586211	0	586470	0
wlp14	**732627**	0	734254	0	734248	0
wlp15	**836570**	0	837423	0	837404	0

but three of the test instances (nos. 13, 14 and 15). Further, it sometimes also fails to eliminate all the store incompatibilities (specifically resulting in one violation each for instances wlp07 and wlp09). The VAM2 heuristic, which uses the 'minimum product of supply cost and store demand' invariant, obtains the best (lowest) cost on four instances and has zero violations on all the test instances. While this is an improvement over VAM1, the best results in terms of the number of lowest costs obtained out of all the test instances are reported for the VAM3 heuristic. The VAM3 heuristic obtains the best results on eight of the test instances, with zero violations overall.

Table 11.2 presents the results obtained by the greedy and randomizedgreedy heuristics. The deterministic heuristics (GDH1 and GDH2) were run once on each test instance, while the randomized greedy heuristics (RGD1 and RDG2) were run a constant $K = 10$ times, and the best solution obtained was reported. Both the greedy and randomized greedy heuristics obtain solutions with zero constraint violations on all the test instances, hence no explicit "violations" columns have been provided in the table. The results show that the SC–SD ratio-based invariant (GDH2) always performs relatively better than the SC–SD product-based invariant (GDH1). Overall, GDH2 reports the lowest-cost solution on five of the test instances. The randomized RGD1 heuristic also reports comparable results, with four lowest-cost solutions overall. The RGD2 heuristic does somewhat better, reporting six best solutions out of 15, once again highlighting the relative

Table 11.2 Results obtained by the greedy and randomizedgreedy heuristics (GDH1, GDH2, RGD1 and RGD2)

Instances	GDH1	GDH2	RGD1 (best)	RGD2 (best)
wlp01	56707	52783	54258	**52563**
wlp02	105555	104968	103992	**102079**
wlp03	141039	138952	**133913**	138182
wlp04	195116	190309	185617	**184094**
wlp05	236453	229373	**225616**	227334
wlp06	260049	246463	**246358**	247625
wlp07	409705	393912	395739	**393007**
wlp08	489820	**462372**	471022	471874
wlp09	560521	537487	544885	**537460**
wlp10	625146	610179	**595678**	605307
wlp11	758855	751103	739081	**736873**
wlp12	803995	**785996**	786289	789069
wlp13	826179	**811393**	819115	815157
wlp14	1084050	**1051989**	1057306	1065850
wlp15	1250741	**1226098**	1240610	1249943

superiority of the SC–SD-ratio-based invariant over the SC–SD product. Individually considered, the randomized heuristics perform better than their corresponding deterministic versions, obtaining comparatively better solutions on all instances. This implies that the greedy store selection approach using either invariant is less effective than using a random order of store selection. Taking the results of the two randomized greedy heuristics together, it is seen that they obtain ten best solutions in total, as against only five best solutions obtained by DGH2 (and no best solutions by DGH1). However, the best performing randomized greedy heuristic (RGD2) is still outperformed by the VAM-based heuristics.

11.6 CONCLUSIONS

Facilities location problems and their variants are wellstudied in optimization theory and find application in various scenarios, such as the effective planning of healthcare facilities, location of transportation terminals, placement of power plants and waste disposal site selection. In this chapter, the broad area of FLPs is briefly reviewed, and a recently introduced variant of the problem called the Capacitated facility location problem with store incompatibilities is studied, in which there are additional constraints that prevent some pairs of stores from being supplied by the same warehouse. The objective of the problem is to minimize the total cost incurred in opening warehouses and allocating warehouses to stores for the supply of goods, while ensuring that there are no conflicting allocations to the same warehouse. The problem is known to be NP-hard, which motivates the use of heuristic solution approaches. The effectiveness of two different heuristic strategies – one based on an effective approximation method (Vogel's method) applied widely to transportation problems and the other consisting of greedy and randomized heuristics – is studied for solving the problem. In both approaches, two simple invariants are used for determining the order in which stores are selected and the basis on which their demands for goods are satisfied from warehouses. The first of these uses the minimum product of supply cost and store demand, while the second uses the ratio of these two parameters. Three Vogel-based heuristics are presented in the chapter, two of which incorporate these invariants. Two greedy heuristics and associated randomized greedy variants are also proposed. The performance of all the heuristics is systematically evaluated on a set of standard benchmark problems. The results show that a randomized approach improves the performance of greedy heuristics for the problem, with further improvements in results reported for the SC–SD ratio-based invariant. The results reported for the Vogel-based heuristics also show that the same invariant proves to be most effective. Since the problem is computationally expensive to solve, these heuristics would prove effective for obtaining good solutions in limited time. The results presented in this work could also be used for developing appropriately hybridized metaheuristics that improve convergence.

REFERENCES

[1] M. Karatas, E. Yakıcı and N. Razi, "Military facility location problems: A brief survey," in *Operations Research for Military Organizations*, IGI Global, 2019, pp. 1–27.

[2] A. Ahmadi-Javid, P. Seyedi and S. S. Syam, A survey of healthcare facility location, *Computers & Operations Research, Elsevier*, vol. 79, pp. 223–263, 2017.

[3] C. Boonmee, M. Arimura and T. Asada, Facility location optimization model for emergency humanitarian logistics, *International Journal of Disaster Risk Reduction*, vol. 24, pp. 485–498, 2017.

[4] C. Tarôco, R. Takahashi and E. G. Carrano, Multiobjective planning of power distribution networks with facility location for distributed generation, *Electric Power Systems Research*, vol. 141, pp. 562–571, 2016.

[5] S. Ilkhanizadeh, M. Golabi, S. Hesami and H. Rjoub, The potential use of drones for tourism in crises: A facility location analysis perspective, *Journal of Risk and Financial Management*, vol. 13, no. 10, p. 246, 2020.

[6] U. Akinc and B. M. Khumawala, An efficient branch and bound algorithm for the capacitated warehouse location problem, *Management Science,* , vol. 23(6), 585–594, 1977.

[7] L.-Y. Wu, X. Zhang and J.-L. Zhang, Capacitated facility location problem with general setup cost, *Computers & Operations Research*, vol. 33, no. 5, pp. 1226–1241, 2006.

[8] J. Beasley, An algorithm for solving large capacitated warehouse location problems, *European Journal of Operational Research*, vol. 33, no. 3, p. 314–325, 1988.

[9] Y. Y. Chen and H. F. Wang, Applying a revised VAM to a multi-level capacitated facility location problem, in *2007 IEEE International Conference on Industrial Engineering and Engineering Management*, pp. 337–341, Singapore, 2007.

[10] D.G. Elson, Site location via mixed-integer programming, *Operational Research Quarterly*, Palgrave Macmillan, vol. 23, pp. 31–43, 1972.

[11] C. Yinka-Banjo and B. Opesemowo, Metaheuristics for solving facility location optimization problem, *Advances in Science, Technology and Engineering Systems Journal*, vol. 3, no. 6, pp. 319–323, 2018.

[12] J. Kennedy and R. Eberhart, Particle swarm optimization, in *Proceedings of ICNN'95- International Conference on Neural Networks*, Perth, 1995.

[13] S. Basu, M. Sharma and P. S. Ghosh, Metaheuristic applications on discrete facility location problems: a survey, *Opsearch*, vol. 52, no. 3, pp. 530–561, 2015.

[14] Z. Ulukan and E. Demircioğlu, A survey of discrete facility location problems, *International Journal of Industrial and Manufacturing Engineering*, vol. 9, no. 7, pp. 2487–2492, 2015.

[15] D. Celik Turkoglu and E. G. Mujde, A comparative survey of service facility location problems, *Annals of Operations Research*, vol. 292, no. 1, pp. 399–468, 2020.

[16] C. Ortiz-Astorquiza, I. Contreras and G. Laporte, Multi-level facility location problems, *European Journal of Operational Research*, vol. 267, no. 3, pp. 791–805, 2018.

[17] M.R.H. Maia, M. Reula, C. Parreño-Torres, P. P. Vuppuluri, A. Plastino, U. S. Souza, S. Ceschia, M. Pavone and A. Schaerf, "Metaheuristic techniques for the capacitated facility location problem with customer incompatibilities", Soft Comput 27, 4685–4698 (2023). https://doi.org/10.1007/s00500-022-07600-z

[18] H. H. Shore, The Transportation Problem and the Vogel Approximation Method, *Decision Sciences*, vol. 1, no. 3–4, pp. 441–457, 2007.

Chapter 12

Novel scheduling heuristics for a truck-and-drone parcel delivery problem

Amrata Gill
Dayalbagh Educational Institute, Agra, India

Prem Prakash Vuppuluri
Dayalbagh Educational Institute (Deemed University), Agra, India

CONTENTS

12.1 INTRODUCTION: BACKGROUND AND DRIVING FORCES

Significant efforts have been made to develop more efficient delivery methods over the past few years, in the wake of the phenomenal growth of the e-commerce industry. One such field that is actively being explored by several global e-commerce and delivery companies is the use of drone-based delivery services. Drones or unmanned aerial vehicles (UAVs) were originally built for military use and often deployed for reconnoitering purposes. However, they currently have wide-ranging applications in several different domains [1]. Drones are being used today for last-mile delivery of consumer goods and services to customers at their doorstep [1], for timely sourcing of medical and humanitarian supplies to disaster-hit areas [2], as well as for transporting pathology samples from hospitals to and from laboratories [3]. Drones equipped with specialized thermal and hyperspectral photographic

DOI: 10.1201/9781003415466-12

equipment and sensors have proved very effective in monitoring large agricultural farms and irrigation systems. They are increasingly being used for tracking cash crops such as corn and rice [3], for detecting pests in agricultural fields [4], for timely identification of diseased crops in fields [5], and for pesticide spraying [6]. A recent application in precision agriculture involved the use of drones for efficiently estimating the water requirements of crops in real time [7]. Drones have also been used to help people follow social-distancing norms [8] and for providing contact-free delivery services. They have been used extensively in transport logistics for better, faster and cost-effective deliveries [9].

However, drone delivery systems have some constraints. For instance, there is an upper limit to the parcel weight, or payload, that can be accommodated on a single run. Similarly, the charge on a drone limits the number of hops and the overall range of delivery. By contrast, traditional truck delivery models, which are based on fossil fuels, have a much longer range. Also, trucks are capable of carrying large and heavy parcels. Overall, however, the traditional truck-based transportation model is time-consuming and incurs high delivery costs. Further, as is the case with most metropolitan road networks, traffic congestion and narrow side roads add to the delivery delays and fuel costs.

Using delivery models that augment the existing truck-based delivery models with drones has the potential to draw on the advantages of both approaches, reduce operational costs and save time. This has been behind the recent trend of using truck-drone combinations for parcel delivery. Specifically, trucks mounted with drones supply goods/parcels to a set of customers/locations, making effective use of drones for some of the items. Several variants of this problem have been studied in recent literature, ranging from single-truck/single-drone to multiple-truck/multiple-drone combinations. In this context, it becomes important that the truck-drone entities follow efficient paths or routes that enable successful delivery of items by both truck and drone, while ensuring that constraints such as time limits, total distance covered, and fuel economy are satisfied. The delivery vehicle(s) typically start from and return to a warehouse or depot and follow a route that covers the customer addresses to which goods need to be delivered.

Many truck-drone delivery routing problems are commonly viewed as variants of the well-known traveling salesman problem (TSP) that occurs in Operations Research, in which a set of cities is provided, along with a starting city, and the objective is to determine the shortest path that goes through each city exactly once before returning to the starting city. A sample instance of the TSP modeled as a graph is shown in Figure 12.1 for eight customers/delivery locations/nodes (marked as red circles). The gray edges indicate (weighted) connectivity between locations, and the black edges represent the shortest path. A possible shortest path that minimizes the distance traveled by the truck is shown, comprising all the black edges. The start and end point of the path is the depot (marked in blue). The TSP is known to be NP-hard

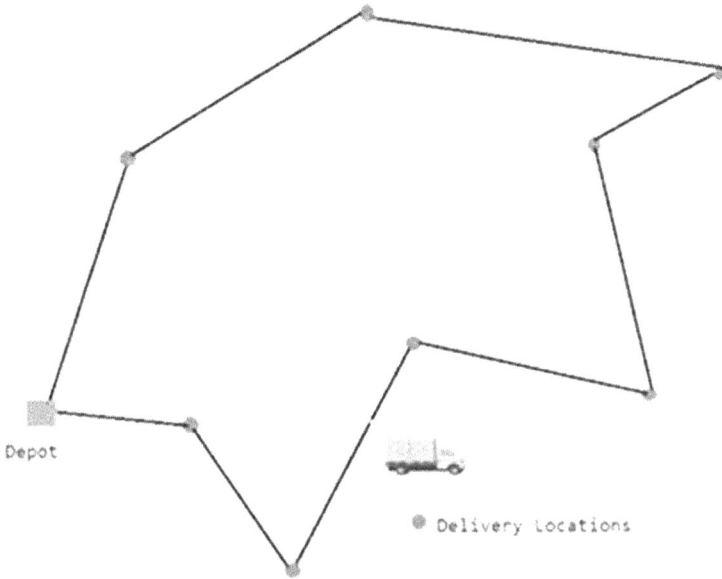

Figure 12.1 TSP example with nine nodes/locations (including start node/depot).

[10], implying that no efficient approach exists for solving the problem to optimality. This characteristic of the problem motivates the exploration of effective heuristic and meta-heuristic strategies for the TSP and its variants.

This article starts with a brief review of various TSP-based truck-drone models and their applications, the object being to provide researchers with a 'big-picture' view of current trends in this area. Thereafter, a novel truck-drone route-planning problem with applications in intra-city parcel delivery is described, and three heuristics for the proposed problem are presented. The performance of the heuristics is evaluated on a set of test instances developed for this work. The remaining sections of the article are organized as follows. A broad overview of the best-known TSP-based truck-drone routing problems, solution strategies developed, and applications is given in Section 12.2. The proposed truck-drone TSP variant is described in Section 12.3, and the TSP heuristics modified for this problem are presented in Section 12.4. Experimental work is described in Section 12.5 and concluding remarks are made in Section 12.6.

12.2 TRAVELING SALESMAN PROBLEM (TSP)-BASED TRUCK-DRONE PARCEL DELIVERY PROBLEMS

Using drones to assist parcel delivery by truck enables the delivery of parcels to be made in a shorter time by mapping packages that would require the truck to travel on slower, more congested roads to the drones, as feasible.

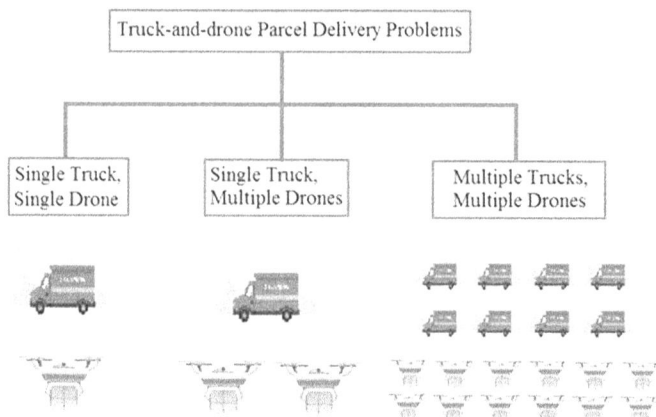

Figure 12.2 Broad variants of the TSP using trucks and drones.

Drones do not require the use of any roads and would therefore be able to deliver goods much faster, and at a lower cost. However, drones are constrained by their relatively small payloads and limited range. Thus, it is important to balance these constraints when planning a transportation solution that involves the use of trucks and drones in tandem.

TSP-based truck-drone parcel delivery problems may be broadly classified based on the number of trucks and drones. In general, a problem may involve the use of one or more trucks, assisted by one or more drones. These problems may be classified as single-truck/single-drone problems, where a single truck is assisted by one drone; as single-truck/multiple-drone problems; or as multiple-truck/multiple-drone problems, as shown in Figure 12.2.

One of the first single-truck/single-drone problems proposed in the literature is the traveling salesman with drone (TSP-D) problem proposed by Agatz, Bouman and Schmidt [11]. Apart from finding a low-cost tour or route for the truck, the TSP-D problem also requires that customers/parcels be partitioned *apriori* between the truck and drone. The authors provide an ILP model and several fast heuristics for the problem. A closely related problem is the 'flying sidekick' TSP (FSTSP) [12], wherein the drone is allowed to take off at one truck delivery location during a tour, deliver a payload to its customer and re-join the truck at another truck delivery location. In this problem, the drone is constrained in that (a) its payload is limited to a single parcel (it may not carry more than one parcel at a time) and (b) it may deliver only a subset of the total set of parcels.

Truck and drone may start from and return to the distribution depot together or independently. For the duration of the tour that they travel together, the drone is based on the truck in order to conserve energy (battery). A certain amount of service time is assumed for operations such as parcel load (before commencement of the tour) and battery recharging/

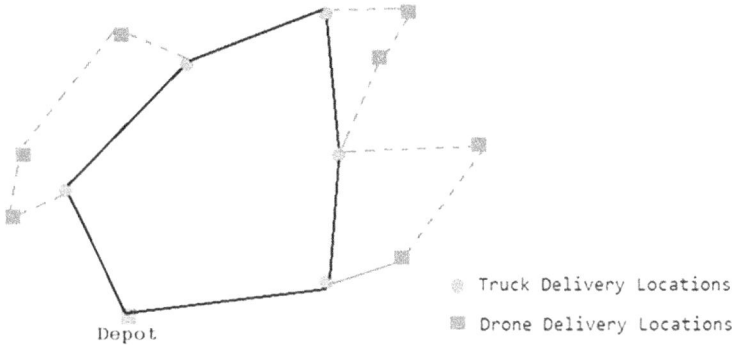

Figure 12.3 FSTSP example with 13 nodes (including depot).

replacement. Travel times for truck and drone are different and unrelated. The time spent hovering by the drone while waiting for the truck to rendez-vous must be minimal, as it also results in depletion of drone battery. Further, neither truck nor drone may visit the same location more than once. The objective is to minimize the completion time (including possible waiting times for drone-truck synchronization) incurred in delivering all the parcels and then returning to the starting point (the warehouse or depot). A sample solution to the FSTSP for a problem instance having 13 nodes (including the depot) is shown in Figure 12.3. In the solution, seven of the nodes are being serviced by drones and the remainder by the truck. Observe how the drone proceeds on a sortie while the truck is servicing one cus-tomer, and then returns at another truck service point/customer. Several extensions to the FSTSP and TSP-D have been proposed. FSTSP is enhanced by Marinelli and Caggiani [13] to allow customer delivery at the node as well as at any point along the route. Chang and Lee [14] describe a different approach that starts by grouping the locations into multiple drone-delivery clusters and then seeks the most efficient delivery route for the truck using the cluster centers.

Poikonen and Golden [15] propose heuristics for a variant of the TSP-D that supports multiple parcels to be carried by a drone, thereby allowing multiple parcel deliveries in a single sortie. The drone is constrained in terms of the total allowable payload, the energy capacity and weight-dependent energy consumption. Drones may launch and land from/at a set of pre-defined locations; however, trucks may not serve customers while the drone is airborne. This last condition assumes that the drone needs to be moni-tored by the truck driver for the duration of its flight, thereby disallowing any other activities for that duration. Boysen et al. [16] consider a variant with a predetermined ordering of nodes that the truck needs to traverse and try to find take-off and landing positions for multiple drones (assuming no range limits) that deliver parcels to groups of customer locations. A special

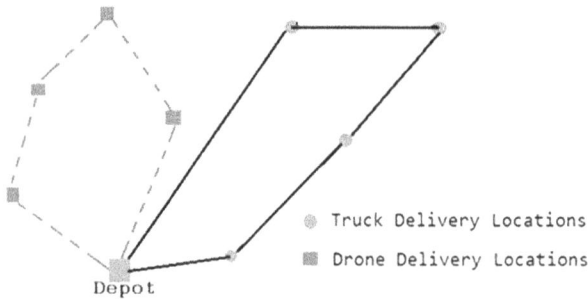

Figure 12.4 PDSTSP example with nine nodes (including depot).

case of this problem is shown to be solvable in polynomial time, but in the general case, it is NP-hard. Two mixed-integer linear programming models for the problem are also proposed. Several meta-heuristic strategies such as GRASP [17], hybrid genetic algorithm [18] and variable neighborhood search [19] have also been applied to the TSP-D and FSTSP problems.

Another related problem is the parallel drone scheduling TSP (PDSTSP), where the drone supports the truck independently by delivering selected parcels (whose size and weight are suitable for drone delivery) directly to a subset of customer locations that lie within the drone's flight range. The remaining parcels are delivered by the truck following a tour that specifies the order in which customers are to be serviced. A sample solution to the PDSTSP for a problem instance having nine nodes (including depot) is shown in Figure 12.4. The figure shows that four of the nodes/customers, which are within the range of the drone, are serviced independently by the drone in a single drone tour. Depending on the drone range and payload capacity, this single tour could in practice devolve into multiple independent tours. The remaining customers are serviced during the truck tour. The problem is introduced by Murray and Chu [12], who also provide two MILP models and greedy heuristics, one each for the FSTSP and PDSTSP. An ant colony-based approach is also proposed for the PDSTSP [20].

An extended variation of the TSP-D is the TSP-mD, which considers the case of multiple drones (>1) on a single truck. An adaptive large-neighborhood search heuristic for this problem is given by Tu et al. [21]. The multiple traveling salesman problem with drones (mTSPD) is a multiple-truck/multiple-drone variant of the problem. This involves multiple tours (one for each truck) and multiple drones performing last-mile deliveries. A MILP formulation for mTSPD along with a two-phase adaptive insertion heuristic and a genetic algorithm are proposed by Kitjacharoenchai et al. [22]. The performance of the GA is compared with that of the standard CPLEX optimizer and shown to obtain solutions in significantly less time.

In the heterogeneous drone-truck routing problem (HDTRP), the role of the truck is to carry (multiple) drones on the tour, stopping temporarily at

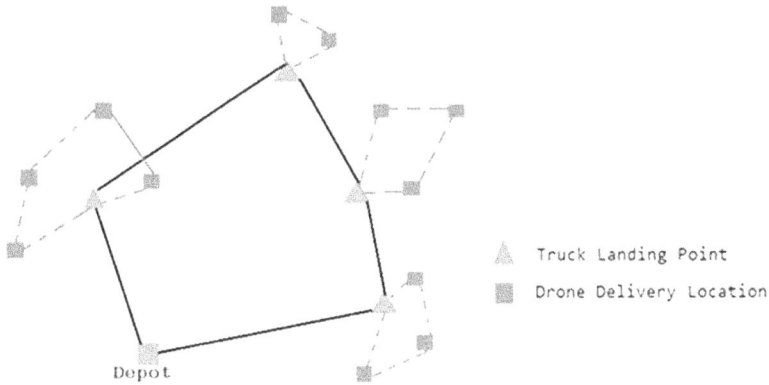

Figure 12.5 HDTRP example with 17 nodes (including depot).

certain landing points from where some of the drones make deliveries and return to the truck. The truck waits until all drones have returned before moving further down the tour. An example solution to the HDTRP on a problem instance having 17 nodes (including depot) is shown in Figure 12.5. A single drone is assumed for this example. The truck stops at four landing points (indicated by the blue triangles), with the drone servicing customer locations near each point. An exact algorithm for solving small instances of the HDTRP is given by Kang and Lee [23].

12.3 CONSTRAINED SINGLE TRUCK-AND-DRONE PARCEL DELIVERY PROBLEM (CSTDP)

This section describes a simplified version of the TSP-D problem that considers delivery points which tend to form small, geographically close clusters. While one of the delivery points in the cluster could be served by the truck, the remaining few are better served by the drone. This variant represents a common package delivery scenario in populated areas where delivery points are distributed across a medium-to-wide geographical area, but small numbers of delivery points often occur relatively close to one another. Given a source warehouse and set of destinations sites, the goal is to construct a tour that enables the delivery truck to start from and return to the source warehouse, while traversing a subset of the delivery points earmarked for delivery by drone. Delivery points covered by the drone would be covered as and when the truck stops to make a delivery at a destination on its tour. Truck-and-drone customers/locations are assumed to be pre-assigned. Further, drone locations are assumed to be within the delivery range of the drone. The objective of the problem is to minimize the total distance required for the truck-and-drone tours.

The main conditions assumed for the problem are as follows:

1. The distance traveled by the drone is computed using the Euclidean norm.
2. The distance traveled by the truck is measured using the Manhattan distance.
3. The drone's vertical take-off and landing times are not considered.
4. For the drone, only the distance traveled is considered, and not the flight time.
5. One drone and one truck operate in tandem.
6. The number of locations serviced by the drone is sufficiently small to warrant a single sortie.
7. Both the truck and drone travel at a constant speed.
8. The truck waits at the starting location of a drone sortie until the drone returns, and subsequently proceeds to the next delivery location; the truck does not travel solely to reunite with the drone.
9. There are no constraints on the airspace traveled by the drone.

12.4 MODIFIED TSP HEURISTICS FOR THE CSTDP

Three heuristics for the traveling salesman problem are adapted for the CSTDP problem in this section. All three are well-known heuristics that are known to obtain good performance for the TSP. In the descriptions that follow in this section, the distances traversed by the drones are quantified using the Euclidean norm (denoted as 'e' in the pseudo-code and discussion), and truck distances (denoted by 'm') are quantified using the Manhattan distance measure.

12.4.1 Modified nearest neighbor (mNN) heuristic

In the modified nearest neighbor (mNN) heuristic, the nearest neighbor is implemented on all the truck locations to determine the route and then the same heuristic is used for determining the drone path for each drone cluster around the truck locations. The key steps of the heuristic are as follows.
Heuristic: mNN

1. Start at the depot w.
2. Mark all the truck nodes as unvisited and the current truck node as visited.
3. Find the truck node that is nearest to the current truck node.
4. For the current truck node, select all drone locations within range of the node:
   a. Mark all the drone locations as unvisited and current location as visited.

b. Find the drone location closest to the current location, mark as visited and add to path via the current location.

c. Repeat step 4(b) until all the drone locations have been added to path.

d. Connect the first location of this subtour to the last visited (drone) location.

5. Repeat steps 3–4 until all the locations are contained in the path. Then link the first and last locations to complete the tour.

12.4.2 Modified nearest insertion (mNI) heuristic

In the modified nearest insertion (mNI) heuristic, the nearest insertion is implemented on all the truck locations to determine the route and then the same heuristic is used for determining the drone path for each drone cluster around the truck locations. The main steps of the heuristic are as follows.

Heuristic: mNI

1. Start at the depot w.
2. Find the truck location i such that m_{wi} is minimal and forms subtour w-i-w.
3. Selection step: Given a subtour T, find a node $k \notin T$, such that distance (k, x) is minimal $\forall x \in T$
4. Insertion step: Find arc(w, j) in the subtour T that minimizes $m_{wk}+m_{kj}-m_{wj}$; introduce k in between nodes w and j:
   a. For node k, select all drone locations within range of the node k.
   b. Starting from truck location k, find the drone location such that e_{kr} is minimal and forms subtour k-r-k.
   c. Selection step: Given a subtour R, find node $r \notin T$ that is closest to any node in the drone tour.
   d. Insertion step: Find the arc(k, s) in the subtour which minimizes $e_{kr}+e_{rs}-e_{ks}$; insert r between k and s.
   e. Go to step 4(c) unless a Hamilton cycle is obtained.
5. Return truck-and-drone paths

12.4.3 Modified minimum spanning tree (mMST) heuristic

In the modified MST heuristic, a minimum spanning tree is constructed using the truck locations, and a pre-order walk of the MST starting from the root is performed, with the root appended to the end of the walk. The well-known Prim's MST algorithm [24] is used for this purpose. The resulting sequence forms the truck tour. A similar approach is used for each drone cluster, using the nearest truck node as both start and end of the subtour. The key steps of this heuristic are as follows.

Heuristic: mMST

1. Mark the warehouse location (w) as the start and end points.
2. Construct MST, taking w as root.
3. List the nodes visited in a pre-order walk of the constructed MST and add w at the end.
4. For each truck location t, mark t as the starting and end location for the corresponding drone locations near t.
   4.1 Construct MST for the drone locations rooted at t.
   4.2 List the drone nodes visited in pre-order walk of the constructed MST and add truck location t at the end.
5. Return truck-and-drone paths.

12.5 EXPERIMENTAL WORK

A random truck-drone dataset generator was built for testing the three constructive heuristics described in Section 12.5. The generator takes three arguments as input: the size of a 2-D grid and upper limits on the number of truck-and-drone locations to be generated, respectively, in the dataset; and returns a 2-D grid populated with appropriate numbers of randomly placed truck-and-drone customer locations. For the experimental work presented in this section, three test sets were generated, each comprising ten test instances. The main characteristics of the test sets, named TSET1, TSET2 and TSET3, are summarized in Table 12.1.

All three heuristics (and also the random dataset generator) are implemented in Python (3.8) and tested on a computing system having Intel core-i5 (2.5 GHz) processor, 8 GB RAM and 512 GB HDD running Windows 11.

The results obtained by the heuristics on the TSET1, TSET2 and TSET3 instances are presented in Tables 12.2–12.4, respectively. In the tables, the 'Inst. No.' filed denotes the instance number, and N_t and N_d denote respectively the number of truck-and-drone locations in the test instance. The tour cost obtained and computation time taken by each heuristic are also reported.

The results for TSET1 (Table 12.2) show that the lowest-cost tours were obtained by the mNI heuristic on all the ten instances. The parcel delivery graph (PDG) obtained by each heuristic on the first instance is also plotted graphically in Figures 12.6–12.8 for the mNN, mNI and mMST heuristics respectively. It is observed that the mNN and mNI heuristics obtain identical

Table 12.1 Summary of test set characteristics

	Grid dimensions	Truck locations	Grid locations
TSET1	500	3–5	10–14
TSET2	100	8–10	17–23
TSET3	500	12–15	30–40

Table 12.2 Tour distances and execution times of the mNN, mNI and mMST heuristics respectively on the TSET1 test instances

Inst. no.	N_t	N_d	mNN heuristic		mNI heuristic		mMST heuristic	
			Cost	Time (s)	Cost	Time (s)	Cost	Time (s)
1	4	13	1752.04	<0.01	**1505.32**	0.04	1985.45	<0.01
2	4	10	2178.83	<0.01	**1966.15**	0.03	2010.41	<0.01
3	4	13	1526.71	<0.01	**1297.49**	0.02	1735.75	<0.01
4	3	10	1735.3	<0.01	**1526.37**	0.03	1649.82	<0.01
5	4	12	2066.68	<0.01	**1850.63**	0.04	1888.73	<0.01
6	5	13	1988.35	<0.01	**1680.1**	0.03	1961.92	<0.01
7	4	14	2137.29	<0.01	**1930.17**	0.04	2351.9	<0.01
8	5	13	1626.28	<0.01	**1483.72**	0.03	1551.17	<0.01
9	3	11	1304.75	<0.01	**1151.81**	0.02	1178.35	<0.01
10	4	12	1823.58	<0.01	**1566.36**	0.02	1598.84	<0.01

Table 12.3 Tour distances and execution times of the mNN, mNI and mMST heuristics respectively on the TSET2 test instances

Inst. no.	N_t	N_d	mNN heuristic		mNI heuristic		mMST heuristic	
			Cost	Time (s)	Cost	Time (s)	Cost	Time (s)
1	9	22	3068.47	<0.01	**2655.16**	0.06	3036.26	<0.01
2	8	24	2726.28	<0.01	**2653.72**	0.05	3267.28	<0.01
3	8	21	2955.79	<0.01	**2701.34**	0.04	2930.03	<0.01
4	9	22	3034.49	<0.01	**2753.85**	0.04	4058.74	<0.01
5	8	18	**2670.12**	<0.01	2739.75	0.05	2878.53	0.02
6	9	22	3338	<0.01	**2976.04**	0.05	3988.62	<0.01
7	9	21	3227.81	<0.01	**3163.11**	0.04	4145.33	0.02
8	8	21	3341.61	<0.01	**3078.85**	0.05	3912.63	<0.01
9	9	27	3217.68	<0.01	**3186.87**	0.06	4037.69	<0.01
10	8	19	3644.19	<0.01	**3022.75**	0.04	3682.08	<0.01

truck tours, but the length of the drone tours constructed using the mNI heuristic is on average shorter. The mMST heuristic obtains the highest tour cost.

The results obtained by the heuristics on the larger test instances (TSET2 and TSET3/Tables 12.3 and 12.4 respectively) also show that the mNI heuristic consistently returns the best results in almost all cases. It outperforms the mMST heuristic on all instances and obtains better tours than the mNN heuristic in 18 out of the 20 test instances, losing out only on instance 2 in TSET2 and instance 5 in TSET3. The PDGs obtained for the first test instance of TSET1 by each heuristic are also depicted graphically in Figures 12.9–12.11. Further, the low computational time reported for the heuristics implies that they scale well for much larger problem sizes.

Table 12.4 Tour distances and execution times of the mNN, mNI and mMST heuristics respectively on the TSET3 test instances

Inst. no.	N_t	N_d	mNN heuristic		mNI heuristic		mMST heuristic	
			Cost	Time (s)	Cost	Time (s)	Cost	Time (s)
1	12	32	3623.24	<0.01	**3183.48**	0.05	6333.99	<0.01
2	14	34	4028.76	<0.01	**3796.47**	0.17	5241.42	0.03
3	14	28	3938.57	<0.01	**3517.79**	0.08	7271.56	<0.01
4	13	36	3744.8	<0.01	**3339.19**	0.06	5420.13	0.01
5	13	28	3577.06	<0.01	**3350.1**	0.06	5044.95	<0.01
6	13	38	4202.75	<0.01	**3970.82**	0.06	5664.11	0.01
7	13	31	3120.15	<0.01	**3017.86**	0.15	6794.7	0.01
8	14	31	4193.21	<0.01	**3806.32**	0.14	5439.85	0.01
9	14	37	4174.95	<0.01	**3997.74**	0.06	5542.72	0.01
10	14	32	3638.26	<0.01	**3526.57**	0.09	6617.72	0.01

Figure 12.6 Truck-and-drone tour obtained by the mNN heuristic on TSET1 instance#1.

Figure 12.7 Truck-and-drone tour obtained by the mNI heuristic on TSET1 instance#1.

Figure 12.8 Truck-and-drone tour obtained by the mMST heuristic on TSET1 instance#1.

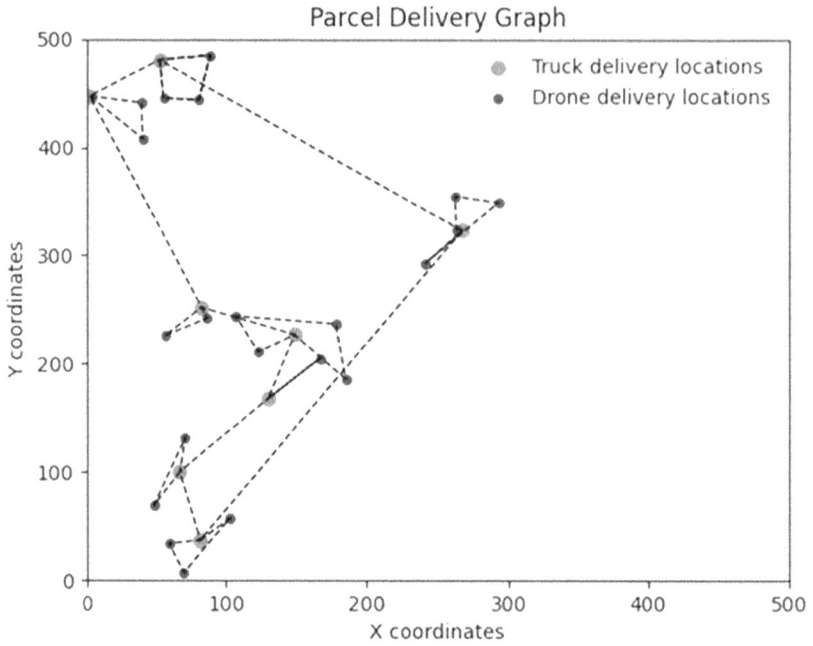

Figure 12.9 Truck-and-drone tour obtained by the mNN heuristic on TSET2 instance#1.

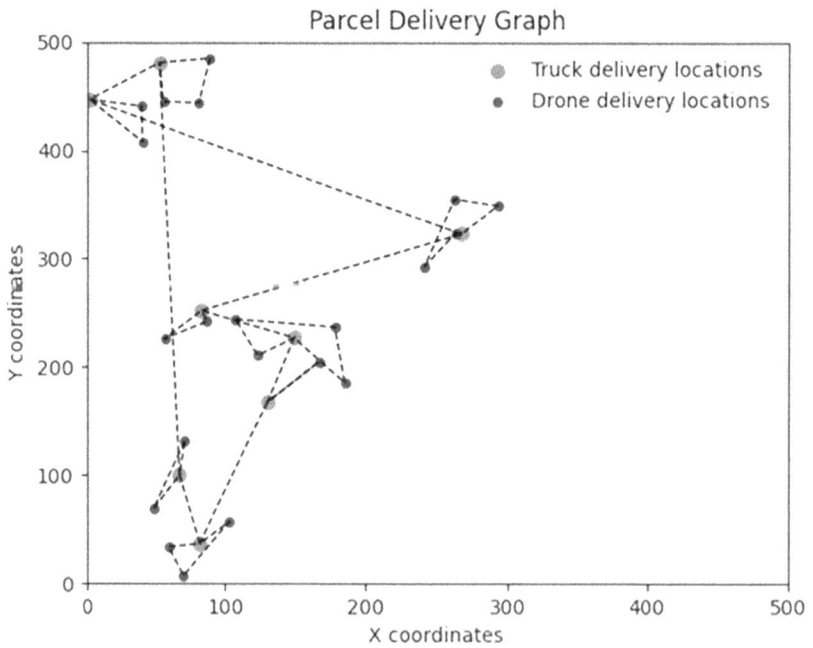

Figure 12.10 Truck-and-drone tour obtained by the mNI heuristic on TSET2 instance#1.

Figure 12.11 Truck-and-drone tour obtained by the mMST heuristic on TSET2 instance#1.

12.6 CONCLUSIONS

Delivery models that combine truck-based delivery models with drones are often modeled as variants of the traveling salesman problem. This chapter introduces and reviews several truck-and-drone problems modeled as TSP variants and discusses the applications of such problems in real-world scenarios. Thereafter, a constrained variant of the single truck-and-drone problem that involves a single truck with a mounted drone for making last-mile deliveries is presented. Three well-known heuristics for the TSP are adapted to this problem, and their performance is compared on a set of 30 specially generated test instances. The results show that the modified nearest insertion heuristic consistently obtains the best results. The TSP is known to be NP-hard, which motivates the development of heuristic and meta-heuristic techniques for solving the problem and its variants. Fast heuristics of the kind studied in this work are applicable for variants that have additional limits on the amount of time available for approximating a tour. The use of such heuristics in hybrid, problem-specific metaheuristics could also lead to significant improvements in their convergence.

REFERENCES

[1] J. T. Bowen Jr, An analysis of Amazon air's network in the United States, *Transportation Journal*, vol. 61, no. 1, pp. 103–117, 2022.

[2] S. M. S. M. Daud, M. Y. P. M. Yusof, C. C. Heo, L. S. Khoo, M. K. C. Singh, M. S. Mahmood and H. Nawawi, Applications of drone in disaster management: A scoping review, *Science & Justice*, vol. 62, no. 1, pp. 30–42, 2022.

[3] A. López, J. Jurado, C. Ogayar and F. Feito, A framework for registering UAV-based imagery for crop-tracking in Precision agriculture, *The International Journal of Applied Earth Observation and Geoinformation*, vol. 97, p. 102274, 2021.

[4] J. Kim, S. Kim, C. Ju and H. Son, Unmanned aerial vehicles in agriculture: A review of perspective of platform, control, and applications, *IEEE Access*, vol. 7, pp. 105100–105115, 2019.

[5] W. Albattah, A. Javed, M. Nawaz, M. Masood and S. Albahli, Artificial intelligence-based drone system for multiclass plant disease detection using an improved efficient convolutional neural network, *Frontiers in Plant Science*, vol. 13, p. 808380, 2022.

[6] A. Hafeez, M. A. Husain, S. P. Singh, A. Chauhan, M. T. Khan, N. Kumar, A. Chauhan and S. K. Soni, Implementation of drone technology for farm monitoring & pesticide spraying: A review, *Information Processing in Agriculture*, 2022. doi: https://doi.org/10.1016/j.inpa.2022.02.002

[7] S. Alexandris, E. Psomiadis, N. Proutsos, P. Philippopoulos, I. Charalampopoulos, G. Kakaletris, E.-M. Papoutsi, S. Vassilakis and A. Paraskevopoulos, Integrating drone technology into an innovative agrometeorological methodology for the precise and real-time estimation of crop water requirements, *Hydrology*, vol. 8, p. 131, 2021.

[8] D. Lu, Drones keep an eye on people failing to social distance, *New Scientist*, vol. 246, no. 3282, p. 10, 2020.

[9] J. Pasha, Z. Elmi, S. Purkayastha, A. M. Fathollahi-Fard, Y. E. Ge, Y. Y. Lau and M. A. Dulebenets, The drone scheduling problem: A systematic state-of-the-art review, *IEEE Transactions on Intelligent Transportation Systems*, vol. 23, no. 9, pp. 14224–14247, 2022.

[10] B. Korte and J. Vygen, Algorithms and combinatorics, in *Combinatorial Optimization: Theory and Algorithms*, Springer Berlin, Heidelberg, 2008, pp. 359–387, doi: 10.1007/3-540-29297-7

[11] N. Agatz, P. Bouman and M. Schmidt, Optimization approaches for the traveling salesman problem with drone, *Transportation Science*, vol. 52, no. 4, pp. 965–981, 2018.

[12] C. C. Murray and A. G. Chu, The flying sidekick traveling salesman problem: Optimization of drone-assisted parcel delivery, *Transportation Research Part C: Emerging Technologies*, vol. 54, pp. 86–109, 2015.

[13] M. Marinelli, L. Caggiani, M. Ottomanelli and M. Dell'Orco, En route truck–drone parcel delivery for optimal vehicle routing strategies, *IET Intelligent Transport Systems*, vol. 12, no. 4, pp. 253–261, 2018.

[14] Y. S. Chang and H. J. Lee, Optimal delivery routing with wider drone-delivery areas along a shorter truck-route, *Expert Systems with Applications*, vol. 104, pp. 307–317, 2018.

[15] S. Poikonen, B. Golden and E. A. Wasil, A branch-and-bound approach to the traveling salesman problem with a drone,*INFORMS Journal on Computing*, vol. 31, no. 2, pp. 335–346, 2019.

[16] N. Boysen, D. Briskorn, S. Fedtke and S. Schwerdfeger, Drone delivery from trucks: Drone scheduling for given truck routes,*Networks*, vol. 72, no. 4, pp. 506–527, 2018.

[17] Q. Ha, Y. Deville, Q. Pham and M. Hà,On the min-cost traveling salesman problem with drone,*Transportation Research Part C: Emerging Technologies*, vol. 86, pp. 597–621, 2018.

[18] Q. Ha, Y. Deville, Q. Pham and M. Hà, A hybrid genetic algorithm for the traveling salesman problem with drone, *Journal of Heuristics*, vol. 26, pp. 219–247, 2020.

[19] J. De Freitas and P. Penna, A variable neighborhood search for flying side-kick traveling salesman problem,*International Transactions in Operational Research*, vol. 27, no. 1, pp. 267–290, 2019.

[20] Q. T. Ding, D. D. Do and M. H. Ha, Ants can solve the parallel drone scheduling traveling salesman problem, in *GECCO '21: Proceedings of the Genetic and Evolutionary Computation Conference*, Lille, France, 2021.

[21] P. Tu, N. Dat and P. Dung, Traveling Salesman Problem with Multiple Drones, in *Proceedings of the Ninth International Symposium on Information and Communication Technology*, Vietnam, 2018.

[22] P. Kitjacharoenchai, M. Ventresca, M. Moshref-Javadi, S. Lee, J. Tanchoco and P. Brunese, Multiple traveling salesman problem with drones: Mathematical model and heuristic approach,*Computers & Industrial Engineering*, vol. 129, pp. 14–30, 2019.

[23] M. Kang and C. Lee, An exact algorithm for heterogeneous drone-truck routing problem,*Transportation Science*, vol. 55, no.5, pp. 1088–1112, 2021.

[24] R. Prim, Shortest connection networks and some generalizations,*Bell System Technical Journal*, vol. 36, no. 6, pp. 1389–1401, 1957.

Chapter 13

A reliable click-fraud detection system for the investigation of fraudulent publishers in online advertising

Lokesh Singh, Deepti Sisodia, Kumar Shashvat and Arshpreet Kaur

Alliance University, Bangalore, India

Prakash Chandra Sharma

Manipal University Jaipur, Jaipur, India

CONTENTS

DOI: 10.1201/9781003415466-13

13.1 INTRODUCTION

Advances in internet technologies have made online advertising an ideal and effective choice for both small- and large-scale businesses to target suitable marketing segments [1]. The key parts of the online advertising ecosystem are:

- *Advertisers.* An advertiser (e.g., eBay) creates an advertisement and generates the revenue for the publisher to display social content (texts, audio content, videos, etc.) on web pages and mobile applications.
- *Publishers.* Publishers provide the platform to display the advertisements on the websites from the advertiser's side.
- *Ad network.* An advertising network (e.g., Value Click) acts as an intermediary/broker between publishers and advertisers. They decide the pricing and fix the revenue for all customer activities such as clicking on advertisements or bidding in an auction.
- *User.* Users are individuals who visit web pages and access advertisements. When the user generates clicks on the displayed ads, they are redirected toward the associated advertiser's webpage [2, 3].

The PPC advertising framework is depicted in Figure 13.1.

Publishers earn revenue using three types of revenue models:

- Cost-per-mille (CPM/PPM): in this model, advertisers pay a fixed amount for every 1000 displays of their message.
- Cost-per-action (CPA/PPA): in this model, revenue is generated according to the desired action. The advertiser pays the cost per action.
- Cost-per-click (CPC/PPC): in this model, revenue is generated for every click performed by the user. The advertisers pay for each click [4].

Of the revenue models, the pay-per-click (PPC) model is prevalent and widely used for online advertisement [5], but it is more vulnerable to suspicious activities because of the large amount of money involved. One such activity is click fraud [6]. Click fraud are fake clicks generated by publishers to earn more revenue without adverts and to mislead ad networks and advertisers [7]. Thus, it is one of the most daunting problems and needs to be addressed [8].

The following challenges are encountered in detecting click fraud:

1. Class imbalance ratio: the number of genuine observations outnumbers the fraudulent observations. The severely skewed nature of the dataset makes the designed model biased towards the outnumbered class.

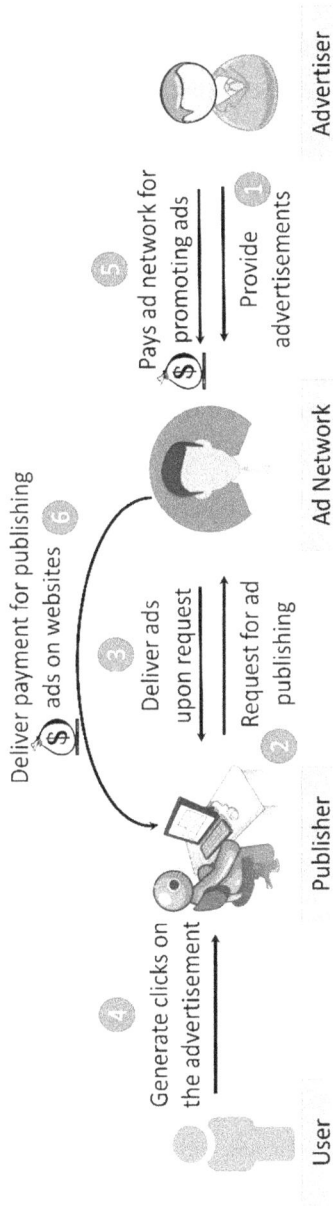

Figure 13.1 PPC advertising ecosystem.

2. Massive data size: millions of clicks per day generated by the publishers or users result in an enormous dataset, which complicates predicting clicks and analyzing the publisher's behavior.
3. Features reported in the literature on click-fraud detection were of low importance, collinear, and redundant.
4. The underlying status of the publisher is not provided. This makes it difficult to investigate which publishers in the CFD dataset should be considered fraudulent, which are genuine, and which should be kept under observation, and hence to design a model for identifying potentially malicious publishers.

The present work overcomes these challenges with a reliable CFD system design that effectively investigates fraudulent conduct of the publishers generating false clicks. The system effectively addresses dataset imbalance with data-level sampling strategies. Since the features reported in the literature on click-fraud detection are of low importance, collinear, and redundant, several filter- and wrapper-based approaches are systematically analyzed in the CFD system. Good, accurate predictive fraud detection features were obtained through majority voting.

The system has investigated a broad range of single and ensemble learning methods to compensate for the lack of comprehensive studies on machine learning methods for detecting fraud in online advertising. Analysis of the feature ranking methods in the conjunction of ensemble learning models led to the discovery of significant optimal features to distinguish between suspicious and normal behaviors. Below are the critical features of the designed framework:

- An efficient click-fraud detection (CFD) system is proposed to effectively investigate fraudulent publishers.
- The issue of class imbalance is resolved by oversampling the fraudulent samples using SMOTE and under-sampling the non-fraudulent instances utilizing RUSBOOST.
- A novel majority voting-based feature selection approach, Hybrid-Manifold Feature Subset Selection (H-MFSS), is proposed to obtain optimal informative features.
- The efficacy of GTB is evaluated on the FDMA2012 mobile advertising user-click dataset for the investigation and classification of illegal publishers.
- To generalize the efficiency of GTB, experiments are also conducted with 12 other conventional individual and ensemble classification models.

The rest of the work is organized as follows. Literature on click-fraud detection for the identification of fraudsters is reported in Section 13.2. Materials and methods for the proposed CFD system are explained in Section 13.3. The classification approaches used in the experiment are discussed in Section 13.4. Section 13.5 examines 10-fold cross-validation, while Section 13.6 covers the major classification assessment methods. The experimental results and discussion are to be found in Section 13.7, while Section 13.8 concludes the work.

13.2 RELATED WORK

Click fraud is an internet crime. While current fraud detection techniques investigate fraudulent actions from multiple perspectives, each has some constraints. The severely skewed nature of the class distribution in the FDMA 2012 dataset for the detection of click fraud has been addressed by Sisodia et al. [9], who evaluated the effect of standard data-level sampling methods, including oversampling, under-sampling, and hybrid sampling techniques, on nine classifiers' performance. Sisodia et al. [10] further addressed the issue by proposing a novel strategy based on prototype selection using k-NN classifier to handle highly skewed class distribution. The QDPSKNN method splits the data into four equal parts and conducts down-sampling to balance the class distribution. It reduces the size of the training set by choosing only closest neighbors as relevant prototypes. In another study, Sisodia et al. [1] resolved the issue of concept drift, in which changes in the publisher's status label make it difficult to identify suspicious behavior. They resolved the issue by proposing a deep CNN-based transfer learning (TL) framework that utilizes distinct pre-trained deep CNN models to extract features, leveraging prior knowledge to avoid learning from scratch. The extraction of optimal and reliable features by the DCNN models (feature extractors) helps identify publisher behavior and classify them as illegal or legitimate from 2-D graphical images using machine learning models.

Sisodia et al. [11] also identified the problem with the unbalanced user-click FDMA dataset. They found that machine learning methods generally fail or provide misleading performance on datasets with a skewed distribution of classes. They used gradient tree boosting (GTB) to eliminate bias from the model, which optimizes loss function and alleviates error at the time of prediction. GTB resolved issues with the FDMA dataset, reducing the bias by integrating several weak learning models. Ways of identifying inflation fraud by analyzing the traffic logs of an ad network are proposed in [12–14].

Nevertheless, low noise attacks by clickbots make detection less efficient. To overcome this problem, Haddadi [6] employed bluff ads to examine fraud partners through the threshold, IP address, and profile matching methods. Metwally et al. designed a model to analyze advertising networks for the detection of single publisher attack [12], hit inflation attack [13], and duplicate click streams [14]. The only constraint observed in this work is that the traffic log of the advertising network remains unavailable to the advertisers. Dave et al. proposed an automated model for detecting and filtering specific click-spam attacks on advertising networks [15, 16]. Li et al. [3] designed a detection system to observe an advertisement delivery process for the detection of fraudulent activity. Despite its advantages, the process is time-consuming and it is hard to detect click fraud in real time. False clicks can be created by clicking bots in the background with a click over a specific link. General browsers are also employed to create malicious clicks. However, the constraint with this is that the traffic created by the general browser might be despoiled through the fraudulent partners and thus transformed into false clicks. The limitation of Asdemir et al.'s proposed model to identify the tradeoff between search engines and advertisers is the overestimation of click fraud because the model is too complex to solve. Berar [17] developed a model which creates the click patterns for every internet advertisement. The model is designed to investigate the status of partners with high precision, but the average precision value obtained is 36.2%, indicating that constraints remain. Berar's model using the random forest ensemble method to examine publisher status respecting their click profile has reduced computational time with an average precision of 49.99% [18]. Perera et al. investigated click statistics using newly created features that combine sets of attributes to analyze behavior patterns of false partners[19].

Most of the attention in the literature has been paid to data-level sampling strategies for balancing the class distribution toward fraudster classification. But there is no comprehensive study of data mining and machine learning methods for fraudster identification. Nor are features comprising crucial information concerning different classes explored. The present work seeks to fill this gap by integrating feature extraction, feature selection, sampling strategies and multiple classifiers into a single framework that effectively predicts the status of a publisher.

13.3 METHODS AND MATERIALS

The framework of the proposed CFD system is demonstrated in Figure 13.2. Since the dataset is highly imbalanced, the first step is to resample the dataset using data-level sampling strategies. Then a variable elimination (feature selection) approach selects an optimal subset of features through majority voting, using filter- and wrapper-based methods. Classification is performed with several individuals and ensemble-based learners.

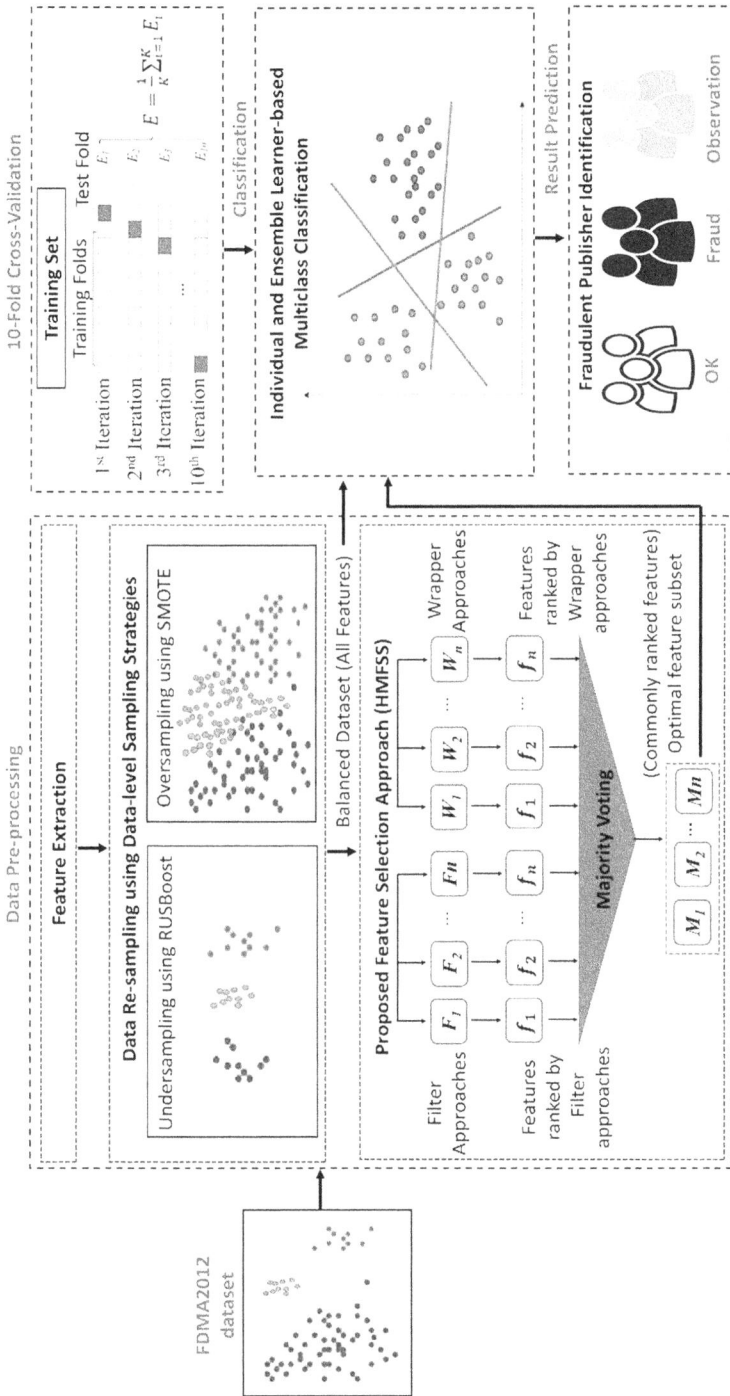

Figure 13.2 Framework of CFD system.

13.3.1 Dataset details

The Fraud Detection in Mobile Advertising (FDMA2012) user-click data sourced by Buzzcity Pvt. Ltd., a mobile advertising company, is used for the experiment in this work [18]. The dataset includes some missing values. The raw data comprises two sets, click and publisher, which fall into three further categories collected at distinct intervals. Details of the dataset can be seen in Figure 13.3, while the click and publisher database samples are listed in Tables 13.1 and 13.2, respectively, indicating missing values in some attributes. The detailed description of click and publisher dataset attributes can be obtained from the following link: www.dnagroup.org/PDF/FDMADataDescription.pdf.

13.3.2 Data pre-processing

As demonstrated in Tables 13.2 and 13.3, the raw data provided comprises noisy patterns and missing values, making click prediction complex. The dataset is therefore pre-processed to remove heterogeneity.

13.3.3 Feature engineering

Extraction of features is the key step in pre-processing; it reduces the dimensionality of the data by extracting more informative and relevant features from the given original data. The heterogeneity of the dataset is thus removed by extracting 103 domain-specific features from the original data for a good data representation.

13.3.4 Data re-sampling using data-level sampling strategies

Classification algorithms generalize well with balanced/equal classes. Thus, they perform poorly when dealing with uneven class proportion due to biasing of the classifier to the outnumbered class. The statistics of the FDMA2012 user-click dataset shown in Figure 13.3 demonstrate severe class skewness in the dataset. The presence of fewer fraudulent publishers than legitimate publishers in the dataset might degrade the performance of numerous machine learning models. Moreover, fewer observations of minority classes might result in false detection of classes [20]. To overcome this issue, we used the potential of two data-level sampling strategies to balance the class distribution of user-click datasets appropriately. Though numerous approaches for handling class imbalance problems are available, in this experiment, RUSBoost is utilized for under-sampling the majority instances, while SMOTE is utilized to oversample the minority samples of the data. The details of these algorithms are as follows.

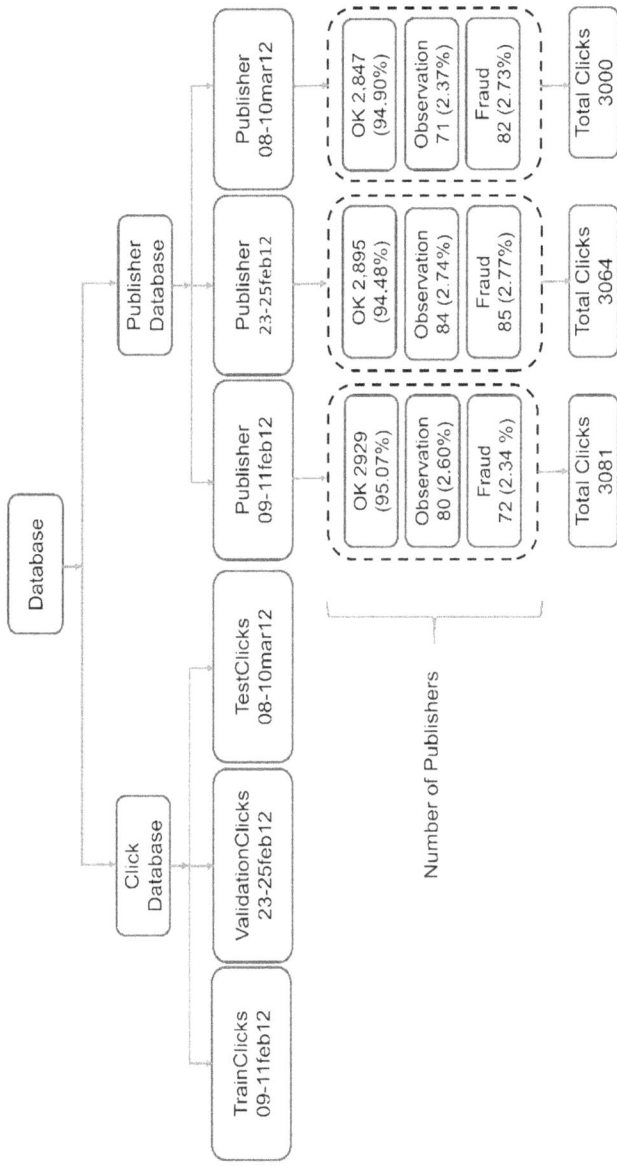

Figure 13.3 Detailed statistics of FDMA2012.

Table 13.1 Statistics of Train_Clicks, Validation_Clicks and Test_Clicks Set

Dataset	Id	Numericip	Deviceua	Pub id	Cam id	Co	Timeat	Cat	Referer
Train clicks	13418094	3358683236	?	8ils9	8ggy8	ar	00:08.0	mc	?
	13418096	3648406743	GT-I9100	8iaxj	8fj2m	ru	00:08.0	ad	24w9x4d25ts00400
	13418100	688845547	Nokia_2700c	8irob	8gpuw	za	00:08.0	co	253mfhuxmmdc4g84
	⋮	⋮	⋮	⋮	⋮	⋮	⋮	⋮	⋮
Validation clicks	3038103	1977772086	Nokia_3110c	8i922	8gt65	in	01:24.0	ad	?
	3038105	2088365076	?	8idhp	8kcce	th	01:24.0	mc	?
	3038107	2945070095	GT-S5360	8ixq4	8k2bq	my	01:24.0	ad	0odkolg7o60gss0g
	⋮	⋮	⋮	⋮	⋮	⋮	⋮	⋮	⋮
Test clicks	9804629	3056950854	GT-S5360	dvry3	duhc8	th	07:37.0	mc	m47q9abkwwc87sz2
	9804630	3384765815	Nokia_E5	du7de	dsfwc	cr	07:37.0	mc	?
	9804631	3384765728	?	dv3va	dsfwc	cr	07:37.0	ad	?
	⋮	⋮	⋮	⋮	⋮	⋮	⋮	⋮	⋮

Table 13.2 Sample click details of the publisher database

Dataset	Published	Bank account	Address	Status
Publisher 9–11 Feb12	8ikt9	0gftajxeyjfccwg8	4fj6vatj8cqocw4k	Fraud
	8iks0	?	2v658mwuvm4g4kow	OK
	8ikpq	?	2z4toc2e5z0gsog0	Observation
	⋮	⋮	⋮	⋮
Publisher 23–25 Feb12	8j0hb	25m0ecmqx000wgw8	35i52az4ufs4ccwc	OK
	8j0g9	?	4cy63ecu4rack8k4	Fraud
	8j0j9	2atn63nfiyzos0ws	2edxb6zg6akgkk44	OK
	⋮	⋮	⋮	⋮
Publisher 8–10 Mar12	dv4bi	?	hxt8v20gxun92dw ikpxbe4rn5	Observation
		il2bojznm9x26g6t	lgit9w6xp008v	
	dv48e	djrcdf4wr	j81tjs8evxxi	OK
	dv48s	?	?	OK
	⋮	⋮	⋮	⋮

Table 13.3 List of top 15 selected features

Final selected features	Feature details	Voting score
102	cntr_sg_percent	6
17	std_spiky_iplong	6
41	evening_std_spiky_referer	4
53	distinct_iplong	4
8	afternoon_avg_spiky_ReAgCnlpCi	5
35	night_std_spiky_referer	3
4	night_avg_spiky_ReAgCnlpCi	6
99	cntr_tr_percent	5
71	std_total_clicks	6
104	cntr_other_percent	5
12	avg_spiky_ReAgCnlp	6
79	first_15_minute_percent	5
86	brand_Apple_percent	6
40	evening_avg_spiky_referer	3
92	avg_per_hour_density	4

13.3.4.1 SMOTE: synthetic minority oversampling technique

SMOTE is an oversampling method that oversamples the proportion of minority classes by creating synthetic samples rather than oversampling with replacements [21, 22]. One of the core advantages of oversampling over under-sampling is that it doesn't result in data loss, as it generates synthetic samples rather than alienating data from the majority class [22]. SMOTE works in the following way. The Fisher-Iris dataset (N) has 110 samples. Among N samples, the two classes, Iris-Versicolor and Iris-Virginica, have 50 samples each, while Iris-Setosa has n = 10 samples. The following steps are required for generating synthetic samples to oversample the minority class:

1. Let one of the minority samples considered (mi) = [5.1, 3.5, 1.4, 0.2]
2. Let the randomly selected nearest neighbor (ra) = [5.2, 3.7, 1.7, 0.1]
3. Computing the positive difference between Step 1 and 2 (di) = [0.1, 0.2, 0.3, 0.1]
4. Random number between 0 and 1 (t) - [0.2, 0.8, 0.3, 0.5]
5. New synthetic minor sample (ms)=$mi * di + t$ → [5.12, 3.66, 1.49, 0.25, 'Iris-Setosa']

The algorithm below represents the pseudo-code of SMOTE oversampling.

ALGORITHM: Smote

Input:

N = Training samples
n = Minority samples
Z = Synthetic samples for every training sample
k = Number of neighbors

Output:

ms = New synthetic samples

Procedure:
Step 1   for i = 1 to n do
Step 2   search k closest centroid neighbor sample
Step 3   for j = Z to 1 do
Step 4   now select one k closest centroid neighbor at random; let's
                       say (ra)
                       $di = ra - mi$
                       t = random number between 0 and 1
                       $ms = mi + di * t$
                       New synthetic samples ← ms
                   end for
               end for

13.3.4.2 RUSBoost

RUSBoost is an under-sampling method that randomly alienates samples from the majority class data. It integrates sampling with boosting to improve the classification performance significantly in case of uneven class distribution [23]. The reason for utilizing RUSBoost is the integration of random under-sampling with the AdaBoost, reducing the time required to design a model. RUS balances the class proportion, and AdaBoost improves the classification performance [24].

Below the algorithm is the pseudo-code of RUSBoost, which is described as follows.

Let D = dataset, x_i = sample point in feature space, y_i = class label in a set of class labels Y, (x_i, y_i) = samples in the dataset, NT = number of iterations while D_P is the desired percentage of total instances to be represented by the minority class, h_t = weak hypothesis and $h(x)$ = outcome of hypothesis. In Step 1, the weight of every sample is set to $1/n$. Here m represents the total number of samples in the training dataset. In Step 2, weak hypotheses are continuously trained, as discussed in Steps 2(i) to 2(viii). In Step 2 (i), random under-sampling is applied to alienate majority class samples until D_P % of the new training dataset D_t belongs to minority observations. Consequently, assign a new weight we_t' to D_t. Now pass D_t and we_t' to base learner and call WeakLearn, which creates the weak hypothesis, as shown in Step 2(iii). After that, evaluate the pseudo-loss for D and we_t' as demonstrated in Step 2(iv). In Step 2(v) weight update parameter α_t is evaluated. The weight division for the next iterative process we_{t+1} is then updated and normalized in Steps 2(vi) and 2(vii). After numerous iterations from Step 2 to 3, Step 4 describes the outcome.

ALGORITHM: Rusboost

Input:
Given training dataset.
$D = \{d_1, d_2, \dots d_N\}$ and minority class $y^m \,\epsilon\, Y$ and $d_i = (x_i, y_i)$ where $x_i \,\epsilon X$ and $y_i \,\epsilon\{-1, +1\}$
NT = number of iterations
D_P = desired percentage of total instances to be represented by the minority class
WeakLearn = Weak learner

Output:
$h(x)$ = final hypothesis

Procedure:
Step 1 Initialize the weight of each sample.

$$w_i(i) = \frac{1}{n} \text{ for all } i \,\epsilon\, \{1 \dots n\}$$

Step 2 (i) for $k = 1$ to NT do
(ii) construct a new training database D_t with weights we_t' by using random under-sampling.

(iii) randomly remove the majority observations until D_p % of the temporary training dataset D_t belongs to minority observations.

(iv) now pass D_t and we_t' to the base learner and call WeakLearn which creates the weak hypothesis.

$$h_t : X * Y \to [0, 1]$$

(v) evaluate pseudo-loss for D and we_t

$$\epsilon_t = \Sigma_{(i,\, y):yi\neq y} we_t(i)(1 - h_t(x_i, y_i) + h_t(x_i, y))$$

(vi) evaluate weight update parameter

$$\alpha_t = \frac{\epsilon_t}{1 - \epsilon_t}$$

(vii) update $e_t : we_{t+1}(i) = we_t(i)\ \alpha_t^{\frac{1}{2}\left(1 + h_t(x_i, y_i) - h_t(x_i, y_i : y \neq y_i)\right)}$

(viii) then normalize we_{t+1}

$$we_{t+1}(i) = \frac{we_{t+1}(i)}{\Sigma_i\, we_{t+1}(i)}$$

Step 3   end for
Step 4   outcome

$$h(x) = argmax_{y \in Y} \sum_{t=1}^{NT} h_t(x, y) \log \frac{1}{\alpha_t}$$

13.3.5 Proposed feature selection approach: hybrid-manifold feature subset selection

The selection of significant features is appropriately deemed a dominant task as it directly affects the model's classification performance. Features obtained after the feature extraction might be irrelevant, increasing the computational cost. The presence of several features during classification not only increases the training time and computational complexity but also tends to overfit the model. Moreover, several irrelevant and noisy features may result in redundant information and affect the accuracy of the predictive model. Therefore, to alienate unwanted and less relevant features, we have designed and evaluated a Hybrid-Manifold Feature Subset Selection (H-MFSS) approach, using the potentials of three existing filter-based approaches and three wrapper-based approaches through majority voting or voting score (VS). VS is the total number of votes gained by an individual feature. The highest VS considered in this experiment is 6, as six existing feature selection approaches are employed in designing the proposed feature selection process. The higher the VS, the higher the relevance of the feature.

The H-MFSS process is demonstrated in Figure 13.2. The two-step working of the designed approach is as follows:

Step 1: **Utilizing Filter and Wrapper Approaches to Obtain Ranking of Features**
  a. *Filter-Based Approach*: Filter-based approaches select the highest voting features evaluated using the heuristic scoring criteria without a learning algorithm [25]. The criteria rank the variables and assign a threshold to filter out irrelevant variables which fall below the threshold [26] and thus are comparably faster and more efficient than wrapper-based methods [27]. GainRatio, Infogain and OneR algorithms are used to obtain highly scored features.
  b. *Wrapper-Based Approach*: Wrapper-based approaches rely on learning algorithms to evaluate the characteristics of selected features. There are two steps: 1) using the search strategy, it seeks the best subset of features; and 2) selected features are then evaluated. These steps are iteratively repeated until a stopping criterion is fulfilled, or the desired classification performance is achieved [28]. Wrapper approaches achieve high classification accuracy but unfortunately lead to high computational complexity due to huge feature space and infinite n number of features, which makes the search unfeasible when the number of features increases [27, 29].

  Features selected by several filter and wrapper methods are shown in Figure 13.4.

Step 2: **Utilizing Majority Voting to Obtain Commonly Ranked Features**
  The feature ranking differs with specific feature selection approaches. Some feature selection approaches provide the feature higher ranking while others provide the same feature lower ranking. Lower-ranked features sometimes degrade performance. Therefore, we used majority voting to identify highly voted and commonly ranked features. The process computes all the votes of each feature provided by the respective feature selection methods and chooses the relevant ranked features. Features with the maximum number of counts are deemed the most relevant and significant. To save computation time, the top 15 most highly voted and commonly ranked features are selected through H-MFSS according to the votes gained by each feature (see Figure 13.4). The details of selected features are represented in Table 13.3, while the voting score gained by selected features is represented in Figure 13.5. The key contribution of feature selection in this experiment is minimum redundancy and maximum relevancy, and this has improved the comprehensibility of models and the predictive precision rate of the classification models employed in this experiment.

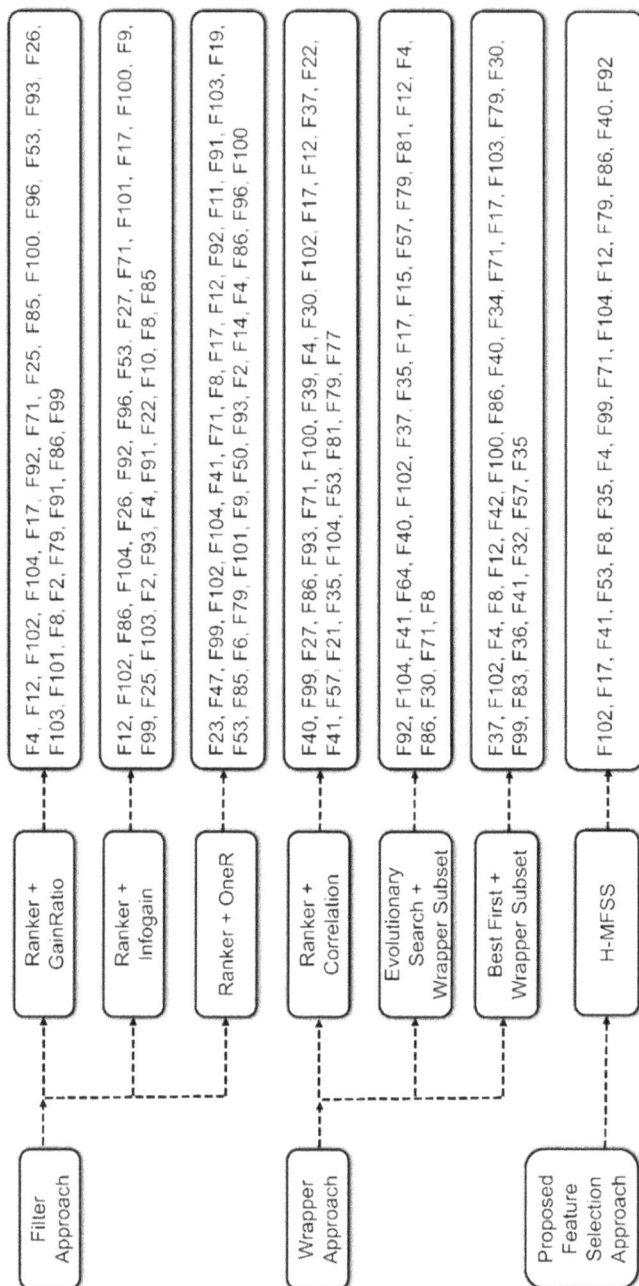

Filter Approach

Ranker + GainRatio
F4, F12, F102, F104, F17, F92, F71, F25, F85, F100, F96, F53, F93, F26, F103, F101, F8, F2, F79, F91, F86, F99

Ranker + Infogain
F12, F102, F86, F104, F26, F92, F96, F53, F27, F71, F101, F17, F100, F9, F99, F25, F103, F2, F93, F4, F91, F22, F10, F8, F85

Ranker + OneR
F23, F47, F99, F102, F104, F41, F71, F8, F17, F12, F92, F11, F91, F103, F19, F53, F85, F6, F79, F101, F9, F50, F93, F2, F14, F4, F86, F96, F100

Wrapper Approach

Ranker + Correlation
F40, F99, F27, F86, F93, F71, F100, F39, F4, F30, F102, F17, F12, F37, F22, F41, F57, F21, F35, F104, F53, F81, F79, F77

Evolutionary Search + Wrapper Subset
F92, F104, F41, F64, F40, F102, F37, F35, F17, F15, F57, F79, F81, F12, F4, F86, F30, F71, F8

Best First + Wrapper Subset
F37, F102, F4, F8, F12, F42, F100, F86, F40, F34, F71, F17, F103, F79, F30, F99, F83, F36, F41, F32, F57, F35

Proposed Feature Selection Approach

H-MFSS
F102, F17, F41, F53, F8, F35, F4, F99, F71, F104, F12, F79, F86, F40, F92

Figure 13.4 Features selected by existing FS methods and proposed approach.

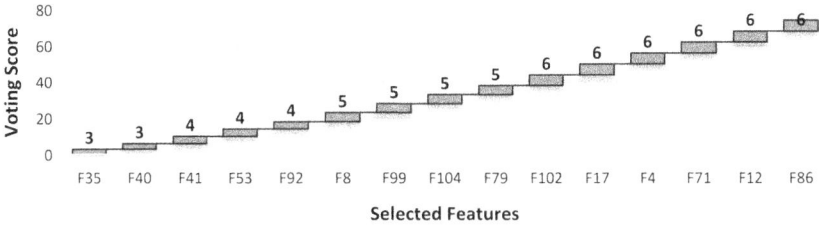

Figure 13.5 Voting score of selected features.

13.4 CLASSIFICATION APPROACHES

This section briefly describes the classification methods used in the experiment.

13.4.1 Gradient boosting machine

The potentials of gradient tree boosting [30, 31], a widely used machine learning model, are utilized to classify fraudulent publishers in this experiment. It builds a predictive model by iteratively integrating several weak learners into a strong learner, using the gradient descent method. The working of GTB is depicted in Figure 13.6.

GTB at the mth thestep would lay a decision tree $h_m(x)$ to pseudo residuals. Suppose the total number of leaves be j_m then the tree splits the input space into several j_m disjoint regions.

$$R_{1m}.......R_mm$$

where constant values are predicted in every region and the output of $h_m(x)$ for input x can be described as the addition:

$$h_m(x) = \sum_{j=1}^{j_m} b_{jm}1R_{jm}(x)$$

where bj_m indicates the predicted value in the region Rj_m.

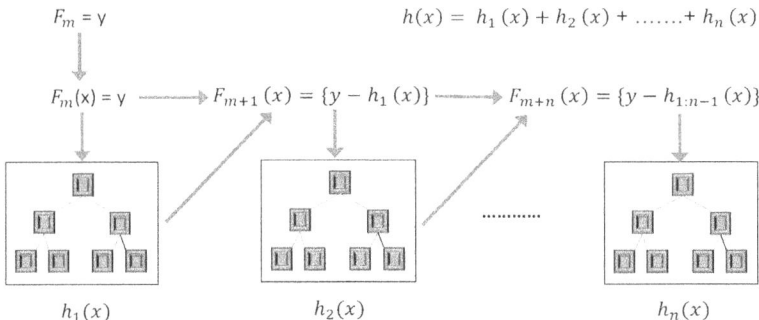

Figure 13.6 Working of gradient tree boosting.

Let γ_m be the selected value employing line search, which is then multiplied by the coefficients bj_m for minimizing the loss function and updating of the model, which is as follows:

$$F_m(x) = F_{m-1}(x) + \gamma_m b_m(x),$$

$$\gamma_m = \arg\ \min_{\gamma} \sum_{i=1}^{n} L\left(y_i, F_{m-1}(x_i) + \gamma b_m(x_i)\right)$$

Rather than selecting a single value γ_m, for the complete tree, Friedman modified this algorithm for selecting the specific optimum values γ_{jm} for every region of the tree. b_{jm} the coefficient is thus alienated from the tree fitting method, and the model is updated as follows:

$$F_m(x) = F_{m-1}(x) + \sum_{j=1}^{jm} \gamma_{jm} 1\, R_{jm}(x)$$

$$\gamma_{jm} = \arg\ \min_{\gamma} \sum_{x_i \in R_{jm}} L\left(y_i, F_{m-1}(x_i) + \gamma\right)$$

The efficiency of GTB is evaluated by experimenting with other state-of-the-art classification approaches, to which we now turn.

Category	Classification method	Description	References
Single Classification Methods	KNN	KNN is a lazy-learner, non-parametric, machine learning classification algorithm which categorizes data using k-nearest neighbors.	[32, 33]
	Decision Trees	The decision tree is a machine learning graph-like structure that designs a training model for the prediction of the class through learning decision rules inferred from the training data.	[34, 35]
	Discriminant Analysis	Discriminant analysis is a machine learning method used for classification which generates distinct classes by distinct Gaussian distributions.	[36, 37]
	Naïve Bayes	Naïve Bayes is a classification approach that follows the Bayes' Theorem principle, which assumes independence among predictors.	[38, 39]
	SVM	Based on the structural risk minimization principle, it carries out classification by searching the optimal hyperplane to maximize the margin between several classes.	[40–43]

Category	Classification method	Description	References
Ensemble Approaches	AdaBoostM2	This approach is the extended version of AdaboostM1 and is suitable for multi-class classification, which utilizes weighted pseudo-loss for n number of observations and k number of classes.	[44–46]
	Random Forest	Random forests build an ensemble of decision trees with the bagging approach to enhance the outcome and the performance of the learning models.	[18, 47, 48]
	LpBoost	Linear programming boost deals with multi-class classification with the aim of maximizing the margin in the training dataset using linear programming.	[49–51]
	TotalBoost	TotalBoost, like LpBoost, generalizes well with the multi-class classification by maximizing the minimum margin using quadratic programming.	[52, 53]

KNN = K-Nearest Neighbor, SVM = Support Vector Machine, GTB = Gradient Tree Boosting

13.5 VALIDATION USING 10-FOLD CROSS-VALIDATION

Cross-validation (CV) is the usual methodology to evaluate the predictive performance of the learners [54, 55]. In this work, a 10-fold CV is employed for the performance evaluation of learning algorithms. 10-fold CV starts with data partitioning into ten equal-sized subsamples/folds [56]. Afterwards, k iterative procedures are performed for training and validation [57, 58]. In every iteration, one-fold among total folds is extracted for testing while the remaining nine are taken for training [59, 60]. The algorithm discussed below represents the pseudo-code for 'k-fold' CV:

ALGORITHM: K-Fold Cross-Validation

Input:
Let D = training set
 k = integer constant

Output:
C = confusion matrix

Procedure:

Step 1   divide the training set D into same-sized subsamples $D_1 .... D_k$
Step 2   for i = 1 to k
       let $S = D/D_i$
       execute the classification algorithm with S as the training set
       test the classifier on D_i to obtain tp_i, fp_i, tn_i and fn_i

Step 3   end for
Step 4   compute $t_p = \sum_i tp_i, f_p = \sum_i fp_i, t_n = \sum_i tn_i, f_n = \sum_i fn_i$
Step 5 C = confusion matrix

13.6 CLASSIFICATION MATRIX AND ASSESSMENT METHODS FOR MULTI-CLASS CLASSIFICATION

The contingency table for binary and multi-class classification is depicted in Figure 13.7, where (a) and (b) represent the confusion matrix for binary class and all three (A, B & C) classes, respectively, where the blue diagonal elements in (b) indicate the actual/true predictions while the remaining represent incorrect predictions. Figures (c), (d) and (e) represent the confusion matrix for A, B, and C classes, respectively. TP_A, TP_B, and TP_C represent classified instances from classes A, B and C. E_{AB} and E_{AC} represent class A instances misclassified as class B or C [61]. False negative, false positive and true negative for classes A, B and C can be calculated as depicted in Table 13.4.

Assessment methods act as an indicator which shows the performance of classification algorithms. The correct interpretation and evaluation of assessment methods is the key to analyzing the performance of the classification model. Though numerous methods are available for the evaluation of classification methods, accuracy and error rate cannot be used as a key measure to handle uneven class distribution. This is because utilizing accuracy as a key measure makes the classifier biased over the outnumbered class rather than the minority class. Since few of the measures are susceptible to imbalanced data, this experiment highlights the evaluation measures used for the classification of multi-class imbalanced data as precision, recall (sensitivity), specificity and geometric mean (G-mean) [20].

Table 13.5 describes the measures for evaluating the multi-class imbalanced dataset.

1. **Positive Prediction Value (PPV) or Precision** is defined as the ratio of accurately classified true instances amid the total predicted instances, as indicated in Eq. 13.1 [61].
2. **Sensitivity/True Positive Rate (TPR)/Hit Rate/Recall** is defined as the measure of correctly classified true samples, indicated in Eq. 13.2.
3. **Specificity/True Negative Rate (TNR)/Inverse Recall** is defined as a measure of accurately classified negative instances, indicated in Eq. 13.3. It is suitable for the evaluation of imbalance in the data due to its dependency on true negative and false positive lying in the similar column of the contingency table [63].
4. **G-mean**: G-mean is the geometric mean of sensitivity and precision [64]. It defines the balance amidst classification performances towards the majority and minority classes [65].

For two class

		Actual Class	
		Positive (P)	Negative (N)
Predicted Class	True (T)	TP	FP
	False (F)	FN	TN
		P = TP+FN	N = FP +TN

(a)

For Multiclass

		Actual Class		
		A	B	C
Predicted Class	A	TP_A	E_{BA}	E_{CA}
	B	E_{AB}	TP_B	E_{CB}
	C	E_{AC}	E_{BC}	TP_C

(b)

For class "A"

		Actual Class		
		A	B	C
Predicted Class	A	TP	FP	FP
	B	FN	TN	FN
	C	FN	FN	TN

(c)

For class "B"

		Actual Class		
		A	B	C
Predicted Class	A	TN	FN	FN
	B	FP	TP	FP
	C	FN	FN	TN

(d)

For class "C"

		Actual Class		
		A	B	C
Predicted Class	A	TN	FN	FN
	B	FN	TN	FN
	C	FP	FP	TP

(e)

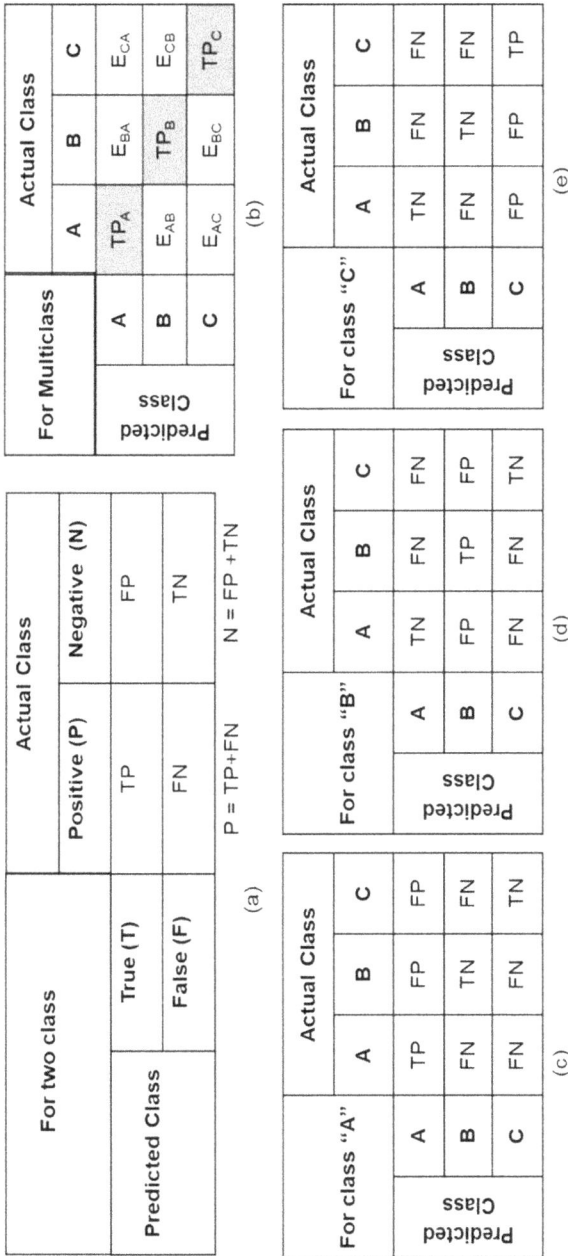

Figure 13.7 Confusion matrix (a) binary and (b–e) multi-class classification.

Table 13.4 Formulations of elements of the multi-class confusion matrix

$FN_A = E_{AB} + E_{AC}$	$FP_A = E_{BA} + E_{CA}$	$TN_A = TP_B + TP_C + E_{CB} + E_{BC}$
$FN_B = E_{BA} + E_{BC}$	$FP_B = E_{AB} + E_{CB}$	$TN_B = TP_A + TP_C + E_{CA} + E_{AC}$
$FN_C = E_{CA} + E_{CB}$	$FP_C = E_{AC} + E_{BC}$	$TN_C = TP_A + TP_B + E_{BA} + E_{AB}$

FN = false negative, FP = false positive, TN = true negative, TP = true positive

Table 13.5 Classification measures for multi-class imbalanced dataset

Classification Measures	References	Equations	Eq. no
Precision	[61, 66]	$PPV = \dfrac{TP}{FP + TP}$	(13.1)
Sensitivity	[62]	$TPR = \dfrac{TP}{FN + TP} = \dfrac{TP}{P}$	(13.2)
Specificity	[63]	$TNR = \dfrac{TN}{FP + TN} = \dfrac{TN}{N}$	(13.3)
G-Mean	[64, 65]	$GM = \sqrt{TPR * TNR} = \sqrt{\dfrac{TP}{TP + FN} * \dfrac{TN}{TN + FP}}$	(13.4)

13.7 RESULTS AND DISCUSSION

Scikit-learn is employed in experiments to evaluate the effectiveness of the designed CFD system to handle challenges that arise in modeling the FDMA 2012 dataset in the fraudster classification. Experimentation among several classifiers has shown GTB to be a significant approach to learning to rank in the user-click dataset. This section evaluates the impact of feature selection and data sampling on the CFD system designed to identify fraudsters.

13.7.1 Impact of sampling

Machine learning methods generally show poor results when dealing with an imbalanced dataset, as they are designed to improve classification performance by alienating error. They therefore do not consider the proportion of classes, resulting in biased and erroneous predictive models due to an imbalance of classes in the dataset. Since the user-click dataset is highly imbalanced and only a few publishers are fraudulent, the model might become biased towards the majority of the classes. This work tackles the problem, using two sampling techniques for under-sampling and oversampling of instances. In oversampling, synthetic samples are generated by oversampling the minority instances, while in under-sampling, the majority instances are reduced.

RUSBoost and SMOTE sampling methods are used to evaluate the effect of sampling in assessing the model's performance. Both methods performed well on the validation dataset. Results in Tables 13.6 and 13.7 demonstrate the impact of sampling using RUSBoost and SMOTE over the learning models. Table 13.6 shows the results analyzed using original data and data sampled through RUSBoost with all features and proposed selected features. In contrast, Table 13.7 shows the results analyzed using actual data and data sampled through SMOTE with all features and proposed chosen features. It can be seen from the results that a balanced dataset gives better classification performance than the imbalanced dataset, where GTB has outperformed the rest of the classifiers. This is illustrated in graph form in Figure 13.8.

13.7.2 Impact of feature selection

The potential of existing filter- and wrapper-based approaches are used in designing this experiment's proposed feature selection strategy. Filter-based approaches are computationally more efficient than wrapper methods, as they evaluate significant features based on data characteristics. Still, selected features are not deemed optimal as filter approaches are independent of learning methods. While wrapper methods use a learning strategy for selecting an optimal feature subset, they guarantee optimal results; an increase in the number of features makes the wrapper method computationally expensive and not feasible. Both methods have pros and cons of their own. The criterion for selecting features usually differs for several feature selection approaches. So, relying on one of the feature selection methods to acquire dominant features may mislead the classifier. Thus, three filter-based and three wrapper-based methods are employed in performing the experiment rather than considering the feature subset obtained from the single method. Utilizing the potentials of conventional feature selection methods, a new Hybrid Manifold-criterion Feature Subset Selection approach is proposed.

The proposed feature selection method has utilized the potentials of three filter-based methods, namely GR (Gain Ratio), IG (Information Gain) and 1R (OneR) and three wrapper-based methods, namely COR (Correlation), ES-WS (Evolutionary Search-Wrapper Subset) and BF-WS (BestFirst-Wrapper Subset) respectively. VS then combines the results of filter- and wrapper-based algorithms to obtain significant and most relevant features where all features have scored respective voting scores. The computation of voting scores is based on the occurrence of a particular feature in the chosen list of all six feature selection methods. If all six feature selection approaches choose six features, those features are deemed most significant

Table 13.6 Performance evaluation of classifiers with all features and selected features on data balanced by RUSBoost

| Methods | Without sampling (all features) | | | | RUSBoost | | | | | | | |
| | | | | | All features | | | | Proposed selected features | | | |
	AP	SE	SP	GM	AP	SE	SP	GM	AP	SE	SP	GM
KNN	43.45	0.42	2.00	0.92	48.92	2.24	2.24	2.24	49.69	2.72	2.39	2.54
LDA	49.63	0.48	1.87	0.95	50.26	2.09	2.1	2.09	51.34	2.36	2.17	2.26
QDA	48.62	0.47	1.33	0.79	44.9	1.56	2.54	1.99	48.88	1.76	2.88	2.25
NB	35.28	0.26	1.29	0.58	39.23	1.26	2.06	1.61	39.69	1.57	2.05	1.79
SVM (L)	35.02	0.34	1.07	0.60	24.45	1.11	2.09	1.52	36.22	1.46	2.03	1.72
SVM (P)	26.89	0.16	1.05	0.41	24.45	1.21	2	1.55	32.27	1.33	2.01	1.63
SVM (RBF)	38.12	0.38	1.22	0.68	36.95	1.47	2.2	1.79	41.3	1.78	2.22	1.98
DT+ CART	47.51	0.5	1.62	0.9	46.93	1.75	2.26	1.98	48.77	1.98	2.38	2.17
ABM2+CART	50.62	0.48	1.95	0.97	56.09	2.01	2.31	2.15	55.03	2.11	2.46	2.27
GTB+CART	**57.05**	**0.58**	**2.56**	**1.21**	**58.87**	**2.65**	**2.84**	**2.74**	**60.67**	**2.94**	**2.97**	**2.95**
RF + CART	**56.02**	**0.55**	**2.21**	**1.10**	**58.38**	**2.32**	**2.63**	**2.47**	**59.89**	**2.64**	**2.93**	**2.78**
LB + CART	56.79	0.55	2.19	1.09	57.38	2.23	2.35	2.28	57.86	2.54	2.42	2.47
TB + CART	52.63	0.49	2.01	0.99	52.87	2.19	2.29	2.23	55.63	2.39	2.44	2.41

Table 13.7 Performance evaluation of classifiers with all features and selected features on data balanced by SMOTE

Methods	Without sampling (all features)				SMOTE								
					All features				Proposed selected features			
	AP	SE	SP	GM	AP	SE	SP	GM	AP	SE	SP	GM
KNN	48.92	0.38	2.43	0.96	50.17	2.34	2.65	2.49	53.8	2.76	2.87	2.81
LDA	54.23	0.43	2.04	0.94	46.67	2.36	2.13	2.24	47.51	2.49	2.23	2.36
QDA	53.29	0.42	1.79	0.87	42.51	2.01	2.16	2.08	56.87	2.08	2.21	2.14
NB	40.57	0.28	1.95	0.74	41.35	1.34	2.11	1.68	35.23	1.89	2.14	2.01
SVM (L)	39.25	0.22	1.88	0.64	43.64	1.32	2.1	1.66	45.26	1.68	2.3	1.97
SVM (P)	31.57	0.15	1.56	0.48	32.25	1.56	1.78	1.67	32.87	1.49	2	1.73
SVM (RBF)	43.28	0.27	1.76	0.69	45.03	1.53	1.99	1.74	47.14	1.66	2.1	1.87
DT+ CART	52.81	0.43	2.23	0.98	55.48	1.32	2.56	1.84	56.64	1.81	2.74	2.23
ABM2+CART	55.9	**0.49**	1.77	0.89	54.83	1.63	1.96	1.79	56.65	1.61	2.3	1.92
GTB+CART	**64.86**	**0.49**	**2.62**	**1.13**	**65.25**	**2.78**	**2.76**	**2.77**	**66.78**	**2.93**	**2.96**	**2.94**
RF + CART	61.13	0.43	2.51	1.04	62.59	2.59	2.83	2.71	62.93	2.82	2.94	2.88
LB + CART	58.75	0.47	2.31	1.04	55.93	2.45	2.46	2.45	58.93	2.68	2.67	2.67
TB + CART	55.54	0.47	2.19	1.01	43.23	2.38	2.26	2.32	55.72	2.59	2.54	2.56

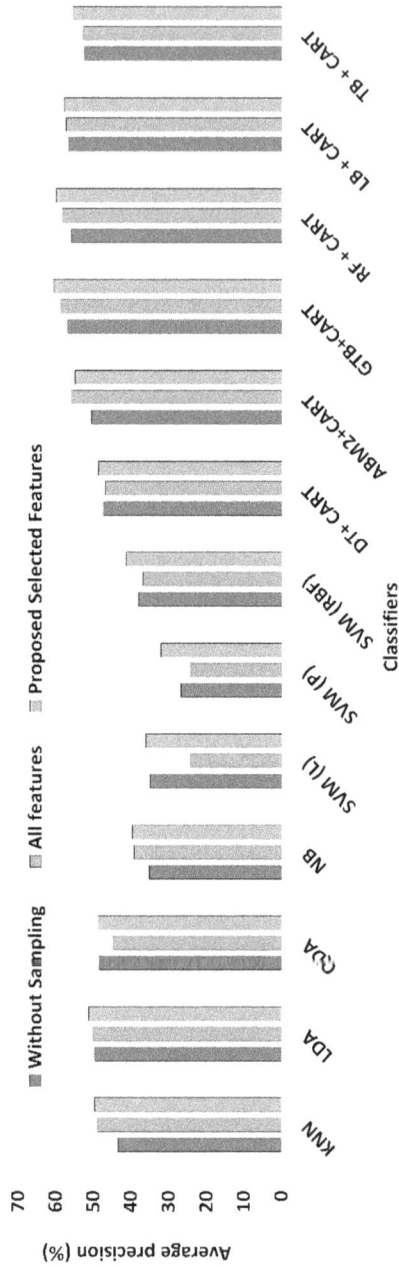

Figure 13.8 Performance comparison of classifiers with all features and proposed selected features on data balanced by RUSBoost.

and voted as '6'. VS = 6 is considered the highest score, while $VS < 3$ is regarded as the lowest score. On this basis, a total of 15 most relevant and significant features are selected as final features by the manifold feature subset selection method to enhance the robustness.

Tables 13.6 and 13.7 show the impact of several existing feature selection methods and the proposed feature selection strategy on the performance of the employed classifiers on the original dataset and dataset balanced by data sampling strategies. The graphical illustration of the same is depicted in Figure 13.9. Results demonstrate that classifiers perform better with selected features utilizing all features. Out of all the methods, Gradient Tree Boosting (GTB) outperforms other classifiers, outperforming other classifiers in terms of average precision, sensitivity, specificity, and geometric mean on the dataset balanced using SMOTE sampling method while simplifying the predictive models.

13.7.3 Performance comparison

The impact of feature selection and sampling strategy on the classifiers over the validation dataset is stated in Tables 13.6 and 13.7. The graphical illustration of the same is demonstrated in Figures 13.8 and 13.9, where classifiers are compared based on the AP score. The result of the GTB classifier has surpassed the results of other classifiers using all features and proposed selected features with and without sampling. The predictive performance of all the classifiers is also evaluated in terms of AP, SE, SP, and GM on the validation dataset. Some of the individual classifiers achieved an average precision of 56% using selected features with sampling, while the results of ensemble classifiers like GTB have generalized the dataset well with an average precision score of 64.86% without sampling (all features), 58.87% with all features and 60.67% with selected proposed features on data balanced by RUSBoost. In contrast, GTB has achieved an average precision score of 65.25% with all features and 66.78% with proposed selected features on data balanced by SMOTE, which is significantly better than results reported by Berar in [17], 36.2%, and Berar in [67], 49.99% on the validation dataset. Random Forest achieved the second-highest classification rate closer to GTB. The key contribution made by GTB in acquiring the highest classification results in terms of average precision, specificity, recall, and G-mean is due to the following strengths:

a. anomaly detection
b. since the construction of trees performed by the GTB is one at a time, the error correction is done by newly constructed trees formed by prior trees, making the model expressive with every added tree.

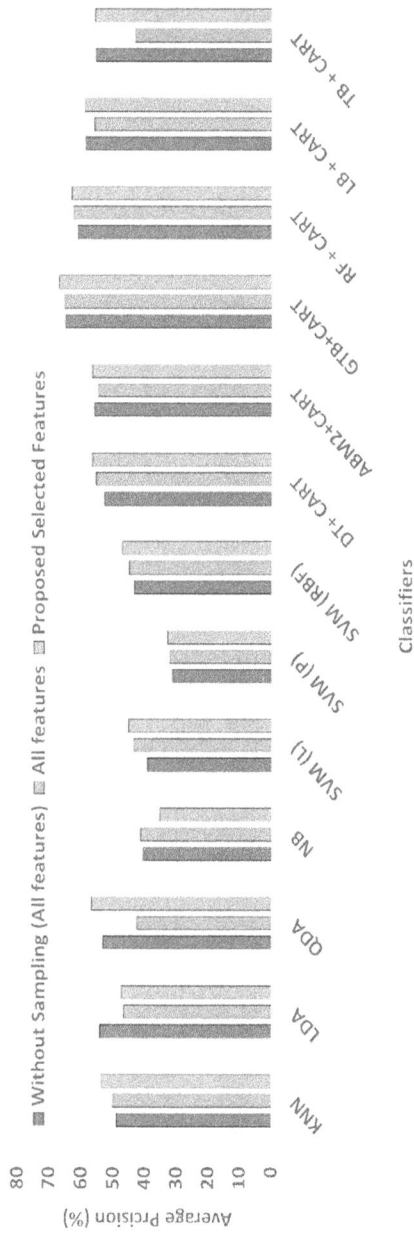

Figure 13.9 Performance comparison of classifiers with all features and proposed selected features on data balanced by SMOTE.

13.8 CONCLUSION

The broad implication of the present research is to identify the suspicious behavior of fake publishers in online advertising. A reliable CFD system for tackling real-world click-fraud detection issues is proposed, utilizing the potential of machine learning methodologies, and representing an invaluable resource for research practitioners. Our work has overcome crucial issues in machine learning research, including the highly unbalanced class distribution, data heterogeneity (combination of categorical and numerical variables), noisy patterns and missing values. The work has proved that exploratory analysis and influential feature extraction made important contributions to effectively detecting click fraud. We have systematically analyzed widely used existing data-level sampling strategies for fair balancing of user-click data. The proposed H-MFSS feature selection approach has led to the generation of effective, optimal features to detect click fraud accurately.

Further, we have investigated the applicability of several distinct individual and ensemble learning models in click-fraud detection tasks. Ensemble models were found to produce significant improvement over individual algorithms. Furthermore, coupling ensemble models with the proposed feature ranking strategy led to the discovery of the essential features that aid in distinguishing fraudulent and normal behaviors of publishers.

ABBREVIATIONS

KNN	K-Nearest Neighbor
LDA	Linear Discriminant Analysis
QDA	Quadratic Discriminant Analysis
NB	Naïve Bayes
SVM (L/P/RBF)	Support Vector Machine (Linear/ Polynomial/Radial Basis Function)
ES-WS	Evolutionary Search + Wrapper Subset
BF-WS	Best First + Wrapper Subset
DT	Decision Tree
ABM2	AdaBoost M2
GTB	Gradient Tree Boosting
RF	Random Forest
LB	LpBoost
AP	Average Precision
SE	Sensitivity
SP	Specificity
GM	Geometric mean
VS	Voting Score
TP	True Positive
FP	False Positive

TN	True Negative
FN	False Negative
GR	Gain Ratio
IG	Information Gain
1R	One R
COR	Correlation
TB	TotalBoost

REFERENCES

[1] D. Sisodia and D. S. Sisodia, "Feature space transformation of user-clicks and deep transfer learning framework for fraudulent publisher detection in online advertising," *Applied Soft Computing*, vol. 125, p. 109142, 2022, doi: 10.1016/j.asoc.2022.109142.

[2] K. Springborn and P. Barford, "Impression Fraud in On-line Advertising via Pay-Per-View Networks," in *22nd USENIX Security Symposium*, 2013, pp. 211–226, doi: 10.4995/Thesis/10251/8685.

[3] Z. Li, K. Zhang, Y. Xie, F. Yu, and X. Wang, "Knowing your enemy: Understanding and detecting malicious web advertising," in *Proceedings of the 2012 ACM conference on Computer and Communications Security*, 2012, pp. 674–686, doi: 10.1145/2382196.2382267.

[4] N. Daswani, C. Mysen, and V. Rao, "Online advertising fraud," *Crimeware: Understanding New Attacks and Defenses*, vol. 40, no. 2, pp. 1–28, 2008, [Online]. Available: http://w.saweis.net/pdfs/crimeware.pdf

[5] D. Sisodia and D. S. Sisodia, "Feature distillation and accumulated selection for automated fraudulent publisher classification from user click data of online advertising," *Data Technologies and Applications*, pp. 1–24, 2022, doi: 10.1108/dta-09-2021-0233.

[6] H. Haddadi, "Fighting online click-fraud using bluff ads," *ACM SIGCOMM Computer Communication Review*, vol. 40, no. 2, pp. 21–25, Apr. 2010.

[7] S. Nagaraja and R. Shah, "Clicktok: Click fraud detection using traffic analysis," in *Proceedings of the 12th Conference on Security and Privacy in Wireless and Mobile Networks*, Miami Florida, 2019, vol. 2, pp. 105–116.

[8] H. Xu, D. Liu, A. Koehl, H. Wang, and A. Stavrou, "Click fraud detection on the advertiser side," in *19th European Symposium on Research in Computer Security, Wroclaw (Poland)*, 2014, pp. 419–438.

[9] D. Sisodia and D. S. Sisodia, "Data sampling strategies for click fraud detection using imbalanced user click data of online advertising: An empirical review," *IETE Technical Review*, pp. 1–10, 2021, doi: 10.1080/02564602.2021.1915892.

[10] D. Sisodia and D. S. Sisodia, "Quad division prototype selection-based k-nearest neighbor classifier for click fraud detection from highly skewed user click dataset," *Engineering Science and Technology, an International Journal*, vol. 28, pp. 1–12, 2022, doi: 10.1016/J.JESTCH.2021.05.015.

[11] D. Sisodia and D. S. Sisodia, "Gradient boosting learning for fraudulent publisher detection in online advertising," *Data Technologies and Applications*, vol. 55, no. 2, pp. 216–232, 2020, doi: 10.1108/DTA-04-2020-0093.

[12] A. Metwally, F. Emekçi, D. Agrawal, and A. El Abbadi, "{SLEUTH}: {S}ingle-pub{L}isher attack d{E}tection {U}sing correla{T}ion {H}unting," *Proceedings of the VLDB Endowment*, vol. 1, no. 2, pp. 1217–1228, 2008, doi: 10.1145/1454159.1454161

[13] A. Metwally, D. Agrawal, and A. El Abbadi, "Detectives: Detecting coalition hit inflation attacks in advertising networks streams," in *Proceedings of the 16th international conference on World Wide Web*, 2007, pp. 241–250, doi: 10.1145/1242572.1242606

[14] A. Metwally, D. Agrawal, and A. El Abbadi, "Duplicate detection in click streams," in *Proceedings of the 14th international conference on World Wide Web - WWW '05*, 2005, p. 12, doi: 10.1145/1060745.1060753.

[15] V. Dave, S. Guha, and Y. Zhang, "Measuring and fingerprinting click-spam in ad networks," in *Proceedings of the ACM SIGCOMM 2012 Conference on Applications, Technologies, Architectures, and Protocols for Computer Communication*, 2012, vol. 42, no. 4, p. 175, doi: 10.1145/2377677.2377715.

[16] V. Dave, S. Guha, and Y. Zhang, "ViceROI: Catching click-spam in search ad networks," in *Proceedings of the 2013 ACM SIGSAC ...*, 2013, pp. 765–776, doi: 10.1145/2508859.2516688.

[17] D. Berrar, "Learning from automatically labeled data: Case study on click fraud prediction," *Knowledge and Information Systems*, vol. 46, no. 2, pp. 477–490, 2016, doi: 10.1007/s10115-015-0827-6.

[18] R. Oentaryo et al., "Detecting click fraud in online advertising: A data mining approach," *Journal of Machine Learning Research*, vol. 15, no. 1, pp. 99–140, 2014, doi: 10.1145/2623330.2623718.

[19] K. S. Perera, B. Neupane, M. A. Faisal, Z. Aung, and W. L. Woon, "A novel ensemble learning-based approach for click fraud detection in mobile advertising," in *Proceedings Mining Intelligence and Knowledge Exploration (MIKE)*, 2013, pp. 370–382, doi: 10.1007/978-3-319-03844-5_38.

[20] S. M. A. Elrahman and A. Abraham, "A review of class imbalance problem," *Network and Innovative Computing*, vol. 1, pp. 332–340, 2013.

[21] N. V. Chawla, K. W. Bowyer, L. O. Hall, and W. P. Kegelmeyer, "SMOTE: Synthetic minority over-sampling technique," *Journal of Artificial Intelligence Research*, vol. 16, pp. 321–357, 2002, doi: 10.1613/jair.953.

[22] B. Santoso, H. Wijayanto, K. A. Notodiputro, and B. Sartono, "Synthetic over sampling methods for handling class imbalanced problems: A review," *IOP Conference Series: Earth and Environmental Science*, 2017, vol. 58, no. 1, p. 12031.

[23] C. Seiffert, T. M. Khoshgoftaar, J. Van Hulse, and A. Napolitano, "RUSBoost: Improving classification performance when training data is skewed," in *Pattern Recognition, 2008. ICPR 2008. 19th International Conference on*, 2008, pp. 1–4.

[24] C. Seiffert, T. M. Khoshgoftaar, J. Van Hulse, and A. Napolitano, "RUSBoost: A hybrid approach to alleviating class imbalance," *IEEE Transactions on Systems, Man, and Cybernetics-Part A: Systems and Humans*, vol. 40, no. 1, pp. 185–197, 2010.

[25] C. Liu, W. Wang, Q. Zhao, X. Shen, and M. Konan, "A new feature selection method based on a validity index of feature subset," *Pattern Recognition Letters*, vol. 92, pp. 1–8, 2017, doi: 10.1016/j.patrec.2017.03.018.

[26] G. Chandrashekar and F. Sahin, "A survey on feature selection methods," *Computers and Electrical Engineering*, vol. 40, no. 1, pp. 16–28, 2014, doi: 10.1016/j.compeleceng.2013.11.024.

[27] N. Hoque, D. K. Bhattacharyya, and J. K. Kalita, "MIFS-ND: A mutual information-based feature selection method," *Expert Systems with Applications*, vol. 41, no. 14, pp. 6371–6385, 2014, doi: 10.1016/j.eswa.2014.04.019.

[28] J. Li et al., "Feature selection: A data perspective," *ACM Computing Surveys (CSUR)*, vol. 50, no. 6, p. 94, 2017.

[29] R. Kohavi and G. H. John, "Wrappers for feature subset selection," *Artificial Intelligence*, vol. 97, no. 1–2, pp. 273–324, 1997, doi: 10.1016/S0004-3702 (97)00043-X.

[30] L. Singh, R. R. Janghel, and S. P. Sahu, "A hybrid feature fusion strategy for early fusion and majority voting for late fusion towards melanocytic skin lesion detection," *International Journal of Imaging Systems and Technology*, no. June, pp. 1–20, 2021, doi: 10.1002/ima.22692.

[31] J. H. Friedman, "Greedy function approximation: A gradient boosting machine," *Annals of Statistics*, vol. 29, no. 5, pp. 1189–1232, 2018.

[32] R. Todeschini, "K-nearest neighbour method: The influence of data transformations and metrics," *Chemometrics and Intelligent Laboratory Systems*, vol. 6, no. 3, pp. 213–220, 1989.

[33] S. Zhang, "KNN-CF approach: Incorporating certainty factor to kNN classification.," *IEEE Intelligent Informatics Bulletin*, vol. 11, no. 1, pp. 24–33, 2010, [Online]. Available: http://www.comp.hkbu.edu.hk/~iib/2010/Dec/article4/iib_vol11no1_article4.pdf

[34] P. E. Utgoff, "Incremental induction of decision trees," *Machine Learning*, vol. 4, no. 2, pp. 161–186, 1989, doi: 10.1023/A:1022699900025.

[35] J. R. Quinlan, "Induction of decision trees," *Machine Learning*, vol. 1, no. 1, pp. 81–106, 1986, doi: 10.1023/A:1022643204877.

[36] J. H. Friedman, "Regularized discriminant analysis," *Journal of the American statistical association*, vol. 84, no. 405, pp. 165–175, 1989.

[37] T. Ramayah, N. H. Ahmad, H. A. Halim, S. Rohaida, M. Zainal, and M. Lo, "Discriminant analysis: An illustrated example," *African Journal of Business Management*, vol. 4, no. 9, pp. 1654–1667, 2010.

[38] N. Friedman, D. Geiger, and M. Goldszmit, "Bayesian network classifiers," *Machine Learning*, vol. 29, pp. 131–163, 1997, doi: 10.1023/a:1007465528199.

[39] N. Friedman and M. Goldszmidt, "Building classifiers using bayesian networks," in *AAAI-96 Preceedings*, 1996, pp. 1277–1284, doi: 10.1.1.30. 4898.

[40] C. Cortes and V. Vapnik, "Support-Vector Networks," *Machine Learning*, vol. 20, no. 3, pp. 273–297, Sep. 1995, doi: 10.1007/BF00994018.

[41] V. N. Vapnik, "Statistical learning theory," *Adaptive and Learning Systems for Signal Processing, Communications and Control*, vol. 2, pp. 1–740, 1998, doi: 10.2307/1271368.

[42] V. N. Vapnik, "The nature of statistical learning theory," *Springer Science & Business Media*, 2013, p. 226.

[43] D. Sisodia, S. K. Shrivastava, and R. C. Jain, "ISVM for face recognition," in *International Conference on Computational Intelligence and Communication Networks, (CICN)*, 2010, pp. 554–559, doi: 10.1109/CICN.2010.109.

[44] R. E. Schapire, "A brief introduction to boosting," *IJCAI International Joint Conference on Artificial Intelligence*, vol. 2, no. 5, pp. 1401–1406, 1999.

[45] Y. Freund, R. Iyer, R. E. Schapire, and Y. Singer, "An efficient boosting algorithm for combining preferences," *Journal of Machine Learning Research*, vol. 4, no. 6, pp. 933–969, 2004, doi: 10.1162/1532443041827916.

[46] R. E. Schapire, Y. Freund, P. Bartlett, and S. Lee, "Boosting the margin: A new explanation for the effectiveness of voting methods," *The Annals of Statistics*, vol. 26, no. 5, p. 1686, 1998.

[47] L. Breiman et al., "Statistical modeling: The two cultures (with comments and a rejoinder by the author)," *Statistical Science*, vol. 16, no. 3, pp. 199–231, 2001.

[48] L. Breiman, "Random forests," *Machine Learning*, vol. 45, no. 1, pp. 5–32, Oct. 2001, doi: 10.1023/A:1010933404324.

[49] Y. K. Fang, "LPBoost with Strong Classifiers *," *International Journal of Computational Intelligence Systems*, vol. 1, no. 2006, pp. 88–100, 2010.

[50] M. K. Warmuth, K. A. Glocer, and S. V. N. Vishwanathan, *Entropy regularized lpboost*. Springer: Berlin Heidelberg, 2008.

[51] K. P. Bennett and J. Shawe-Taylor, "Linear programming boosting via column generation," *Machine Learning*, vol. 46, pp. 225–254, 2002.

[52] G. Lemaitre and M. Radojevic, "Directed reading: Boosting algorithms," *Heriot-Watt University, Universitat de Girona, Universite de Bourgogne*, pp. 1–13, 2009.

[53] M. K. Warmuth, J. Liao, and G. Rätsch, "Totally corrective boosting algorithms that maximize the margin," in *Proceedings of the 23rd International Conference on Machine Learning – ICML '06*, no. 1999, pp. 1001–1008, 2006, doi: 10.1145/1143844.1143970.

[54] Y. Zhang and Y. Yang, "Cross-validation for selecting a model selection procedure," *Journal of Econometrics*, vol. 187, no. 1, pp. 95–112, 2015.

[55] L. Singh, R. R. Janghel, and S. P. Sahu, "SLICACO: An automated novel hybrid approach for dermatoscopic melanocytic skin lesion segmentation," *International Journal of Imaging Systems and Technology*, pp. 1–17, 2021, doi: 10.1002/ima.22591.

[56] L. Singh, R. R. Janghel, and S. P. Sahu, "An Empirical Review on Evaluating the Impact of Image Segmentation on the Classification Performance for Skin Lesion Detection," *IETE Technical Review (Institution of Electronics and Telecommunication Engineers, India)*, , no. May, 2022, doi: 10.1080/02564602.2022.2068681.

[57] L. Singh, R. R. Janghel, and S. P. Sahu, "A deep learning-based transfer learning framework for the early detection and classification of dermoscopic images of Melanoma," *Biomedical and Pharmacology Journal*, vol. 14, no. 3, pp. 1231–1247, 2021, doi: 10.13005/bpj/2225.

[58] L. Singh, R. R. Janghel, and S. P. Sahu, "A boosting-based transfer learning method to address absolute rarity in skin lesion datasets and prevent weight drift for melanoma detection," *Data Technologies and Applications*, 2022, doi: 10.1108/DTA-10-2021-0296.

[59] P. Refaeilzadeh, L. Tang, and H. Liu, "Cross-validation," in *Encyclopedia of Database Systems*, Springer, 2009, pp. 532–538.

[60] L. Singh, R. R. Janghel, and S. P. Sahu, "TrCSVM: A novel approach for the classification of melanoma skin cancer using transfer learning," *Data Technologies and Applications*, pp. 1–18, 2020, doi: 10.1108/DTA-06-2020-0126.

[61] A. Tharwat, "Classification assessment methods," *Applied Computing and Informatics*, pp. 1–13, 2018, doi: 10.1016/j.aci.2018.08.003.

[62] M. Sokolova and G. Lapalme, "A systematic analysis of performance measures for classification tasks," *Information Processing and Management*, vol. 45, no. 4, pp. 427–437, 2009, doi: 10.1016/j.ipm.2009.03.002.

[63] M. Hossin and M. N. Sulaiman, "Review on evaluation metrics for data classification evaluations," *International Journal of Data Mining & Knowledge Management Process (IJDKP)*, vol. 5, no. 2, pp. 1–11, 2015, doi: 10.5121/ijdkp. 2015.5201.

[64] R. P. Indola and N. F. F. Ebecken, "On extending F-measure and G-mean metrics to multi-class problems," in *Sixth International Conference on Data Mining, Text Mining and Their Business Applications, UK*, vol. 35, pp. 25–34, 2005, ISSN 1743-3517.

[65] M. Bekkar, H. K. Djemaa, and T. A. Alitouche, "Evaluation measures for models assessment over imbalanced data sets," *Journal of Information Engineering and Applications*, vol. 3, no. 10, pp. 27–38, 2013, [Online]. Available: http://www.iiste.org/Journals/index.php/JIEA/article/view/7633

[66] D. M. W. Powers, "Evaluation: From precision, recall and F-measure to Roc, informedness, markedness & correlation," *Journal of Machine Learning Technologies*, vol. 2, no. 1, pp. 37–63, 2011, doi: 10.1.1.214.9232.

[67] D. Berrar, "Random forests for the detection of click fraud in online mobile advertising," in *Proceedings of 2012 International Workshop on Fraud Detection in Mobile Advertising (FDMA), Singapore*, 2012, pp. 1–10, [Online]. Available: http://berrar.com/resources/Berrar_FDMA2012.pdf

Chapter 14

Crypto-currency analytics and price prediction

A survey

Vikas Bajpai, Anukriti Bansal, Ashray Mittal and Pratyush Chahar

The LNM Institute of Information Technology, Jaipur, India

CONTENTS

14.1 INTRODUCTION

One of the major fields to be impacted by today's technology boom is the finance industry. The year 2006 saw the introduction of the first digital currency, Bitcoin [1], which has been followed into the market by many similar currencies known as crypto-currencies. Traditional currency systems were monitored by third-party organizations like central or national banks, whereas Bitcoin is a p2p electronic cash system, that is, it enables person-to-person money transfer without the involvement of a mediator to verify

DOI: 10.1201/9781003415466-14

the transaction. The technology behind the formation of crypto-currencies was blockchain, a distributed database which maintains and distributes all types of transactions and digital events. It is a peer-to-peer network that acts as a shared ledger where all transactions are confirmed by the network and cannot be falsified. Bitcoin prices are predicted and estimated for their intrinsic value and for investment purposes. These prices are affected by various microblogging platforms, such as Twitter and Google Trends, that help to identify people's attitudes.

This chapter describes the various algorithms and methodologies used in price prediction. The remainder of the chapter is organized as follows. Section 14.2 examines the history of Bitcoin, and Section 14.3 explains blockchain, the technology behind most crypto-currencies. Section 14.4 discusses the impact of Twitter on crypto-currency prices. Sections 14.5 explores methodologies previously used to predict prices; the advantages and disadvantages of the various methods and algorithms used by previous researchers are considered in Sections 14.6 and 14.7, respectively. Section 14.8 discusses the importance of combining new technologies with experience, and Section 14.9 looks at the accuracy of previously used prediction algorithms. Section 14.10 discusses ethical aspects and the dangers and implications of data misuse, and Section 14.11 offers some conclusions.

14.2 BITCOIN

The most popular and widely used of the large number of crypto-currencies in today's market is Bitcoin. Unlike traditional currencies, Bitcoin's value is based on computer complexity rather than on real goods; thus it is connected with cryptography, software engineering, and economics. The majority of world currencies are governed by governments, either directly or indirectly (i.e., through a central bank). In both circumstances, it is a government's goals and policies that steer and manage its currency. While this holds true of central banks, they also have more access control and enjoy a semi-independent relationship with the government. The role of the central bank is to assist the government in achieving its objectives in areas such as economic stability, growth, and currency value stability. Bitcoin, however, has no central authority to directly influence price or supply, which eliminates the intermediary that traditional monetary systems rely on, namely the central bank and banking system. Participating in transaction calculations is the only way to increase Bitcoin supply, which in turn pays for the infrastructure. The currency's monetary value is determined by the same factors that affect the value of a fiat currency [31].

Bitcoin is a decentralized peer-to-peer (p2p) network which allows funds to be moved straight from one party to another without the requirement for a third party to validate the transaction. The ledger, a key aspect of the decentralized design, is a shared database that all nodes have a copy of.

This ledger keeps track of all previous Bitcoin transactions as well as current Bitcoin owners. The database is organized into chronological transaction blocks. A new block is formed by aggregating recent transactions and then cryptographically linking it to previous blocks, forming a chain of blocks known as a blockchain. This design makes it difficult to criticize or change a previous block in the chain, making it secure and transparent. Crypto-currencies create an environment in the owners' minds where no central authority can access their money, and the owner is the sole governing body of that currency [47].

14.3 BLOCKCHAIN

Satoshi Nakamoto's (2005) paper, "Bitcoin: A p2p electronic cash system", was the first to popularize blockchain [1]. The blockchain is a distributed database that records and distributes every single transaction and digital occurrence. The bulk of the system's members double-check each transaction. Applications of blockchain technology go far beyond peer-to-peer payment systems to include IOT applications, distributed storage systems, healthcare, and other applications, because it provides security, privacy, and a distributed ledger [3]. Owing to blockchain's vast range of applications, many new blockchains have emerged and there are now around 1,600 crypto-currencies [33].

Crypto-currencies provide the finance for systems to run and authenticate the blockchain, as well as the electricity required to do so. Crypto-currencies will become more popular as the use of blockchains grows. This is their main intrinsic worth, but their value depends on a variety of conditions. Gaining a better grasp of what can drive price shifts adds value [1].

14.4 TWITTER

Microblogging is a type of social media that lets everyone communicate their opinions in smaller, more regular updates than blogs. The social networking platform Twitter began in July 2006 as a microblogging tool. Twitter enables users to send public messages, known as "tweets", that were originally limited to 140 characters in length, increased to a 280-character limit in November 2017. End users can also use "hashtags", which consist of the "#" symbol followed by a string of characters, in their tweets. The advantage of this is that it determines the theme of a tweet and makes it available for searches [6].

Twitter's popularity has skyrocketed since its inception in 2006. The company's power and reach was powerfully illustrated when it was an image shared on Twitter, rather than traditional news media, that broke the story of the US Airways flight that ditched in the Hudson River on January 15,

2009 [6]. As of now, Twitter has a total of 1.3 billion accounts and 330 million active monthly users; 83 percent of the world's leaders own a Twitter account. Bots account for roughly 23 million of Twitter's active users, and 500 million tweets are sent every day.

Twitter is therefore a powerful source of public opinion on practically any topic [7], and changing sentiments can be tracked by following the timelines of tweets. Twitter is an invaluable guide for gathering text data on a topic such as crypto-currency price connections [29].

Although Twitter is losing users, and some experts believe the service has reached the limits of its user base, the company's 97% gain in advertising sales is positive. Also, these advertising strategies effect major fluctuations in the price of different company stocks and crypto-currencies. Twitter is thus an interactive tool for not only gaining global insights but also predicting future prices. This tool acts as a sample space or dataset that allows us to work with internet data and uses various kinds of data prediction methods already available. Some of these price prediction algorithms used are described in the next section. [2]

14.5 MAJOR ALGORITHMS AND WAYS TO PREDICT CRYPTO-PRICES

Plenty of work has already been done in this field. This section presents some of the major methodologies and algorithms used.

14.5.1 Sentiment analysis

Natural language processing (NLP) is used for the analysis and classification of sentiment from Twitter and Google Trends reviews. Twitter generates a large amount of sentiment-rich data in the form of posts and hashtags in comments. The sentiments extracted from comments provide indicators for many different purposes. Sentiments are categorized broadly as either positive or negative. In this modern era, people rely on online content for many decisions [14]. For example, when we want to read any research paper, we usually look up the reviews and the number of downloads. As content generated over the internet is too huge to analyze, there is a requirement for various sentiment analysis approaches. Many tasks are included in sentiment analysis, including sentiment extraction, sentiment classification, and subjectivity categorization. Sentiment analysis can be used to identify whether tweets gathered from Twitter are favorable or negative when it comes to crypto-currencies. There are a number of approaches:

Lexicon-based approaches.
To determine polarity, these techniques use sentiment dictionaries with opinion terms and connect them to the data. Sentiment points are

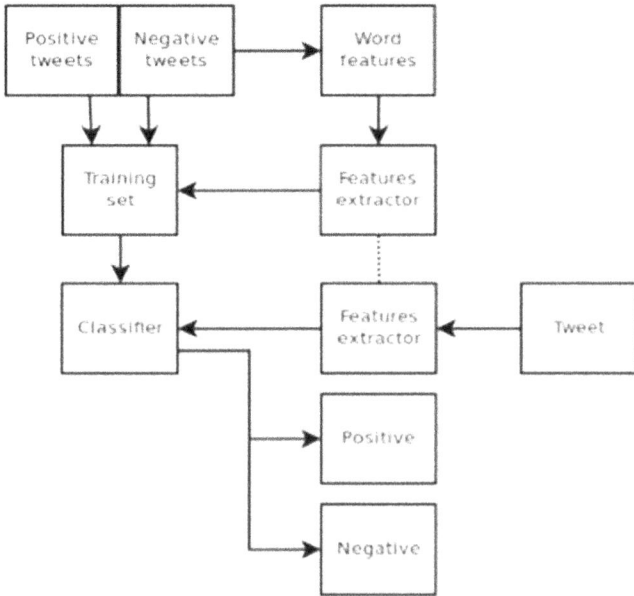

Figure 14.1 A general sentiment analysis model.

assigned to the words which describe the words as positive or negative. These approaches are dependent on recognized phrases. Approaches based on lexicons operate on external directories. There are two types of lexicon-based approach:

Dictionary based: Collected terms are classified using a dictionary. The set expands with the search for more synonyms and antonyms.

Corpus based: Corpus-based classification sends dictionaries linked to a particular domain. The dictionaries provided are created from a group of terms using either semantic or statistical techniques.

Figure 14.1 shows a general model for sentiment analysis.

14.5.2 Naïve Bayes

Naïve Bayes is a classification technique based on Bayes' Theorem that works on the concept of probability. It differentiates word list contents and assigns them to the appropriate class. It is predicated on predictor independence: in simple terms this classifier believes that the presence of features is not co-related in any way.

The class is defined by C^* which is given by:

$$C^* = \arg \arg \mathrm{mac}_c P_{\mathrm{NB}}(c \mid d)$$

$$P_{NB}(d) = \frac{(P(c)) \sum_{i=1}^{m} p(f \mid c) n_{i(d)}}{P(d)}$$

where:
   f: feature
   $n_{i(d)}$: feature count
   m: number of features
   Maximum likelihood estimates are used to calculate probabilities like $P(c)$
   and $P(f|c)$. To implement naïve Bayes, the NLTK Python library is used.

14.5.3 Support vector machine

Support vector machine (SVM) is a linear model which can help solve real-world classification and regression problems, both linear and non-linear. One of the features of SVM is that a line or a hyperplane that divides the data into classes is produced by the algorithm. SVM is used to analyze data by defining the decision boundaries and using kernels to perform computations [30]. The data that is provided as input consists of two sets of m-dimensional vectors. The distance between the divided classes is defined by the margin of the classifier and the decision is based on the margin. Increasing the margin minimizes the number of borderline decisions. SVM aids the accurate recognition of components as well as classification and regression, both of which are helpful for learning.

14.5.4 Decision tree

Decision tree analysis creates a tree-shaped diagram for performing a task or a statistical probability analysis. This technique is used to solve difficult issues. The decision tree's branches each represent potential outcomes. For calculations, the population is divided into at least two groups. The test on the features is determined by the internal nodes, the result is portrayed by the branch and the leaves are the decisions made after successive processing [50]. The decision tree works as follows:

a. At the tree's root, place the dataset's best feature.
b. Divide the dataset into a training and test set. Subsets must be created so that each subset has data with the same feature attributes.
c. The procedures above should be performed for each subgroup until we have a leaf in the tree.

14.5.5 Bayesian regression

Using probability distributions rather than point values, we define linear regression from a Bayesian perspective. Instead of attempting to isolate the one "best" value for each model parameter, Bayesian linear regression seeks to identify the posterior distribution of those values.

The algorithm for Bayesian regression is as follows:

a. The data should be broken into all possible consecutive intervals. Then k-means clustering is used further to retrieve some cluster centers for every kind of interval size that we know. The sample entropy method is used to reduce them to some of the varied and most useful clusters.

b. The next step is to calculate the corresponding weights of features discovered in this regression approach using the second set of prices. The working of regression is as follows:
   - At time t, we have to calculate 3 vectors of previous prices of dissimilar time intervals (180s, 360s and 720s).
   - Then the similarity between the vectors is calculated and our 20 best k-means patterns with acknowledged price jumps for each time interval is used to determine the "probabilistic price change" dp.
   - The weights w_i iare then calculated using a differential evolution optimization method for every available feature.

c. The algorithm is evaluated using the third set of prices, which are obtained by using the same Bayesian regression to calculate characteristics and joining them with the attached weights established in step b [16].

14.5.6 ARIMA

Autoregressive integrated moving average (ARIMA) is a model which analyzes data based on statistics and uses time-series data to comprehend the data set. It is effective for understanding and predicting time-series data. Its use in predicting unemployment rates, weather forecasts and many such problems has been well documented. The ARIMA model assumes stationarity, therefore we need to determine the three parameters p, d, q.

p: autoregressive component
q: moving average
d: degree of differencing

According to Dickey-Fuller and KPSS, tests d should be greater than zero (d > 0) for stationarity [19].

14.6 ADVANTAGES OF DIFFERENT ALGORITHMS

Each of the methods for predicting crypto-currency prices has its own advantages. The support vector machine performs really well where the number of dimensions is greater than the number of samples. This algorithm is powerful in spaces that are high-dimensional [30].

Deep learning algorithms used in neural networks such as RNN and CNN have some advantages: CNN performs weight sharing and can be used to calculate large data set prices very easily [32]. Sentiment analysis uses the vast amount of data available on microblogging sites such as Twitter as a valuable source of public opinion [14].

The major benefit of ARIMA is that it uses time-series values, a useful functionality if we are forecasting a huge number of time series. It also addresses a difficulty that multivariate models occasionally face. Instead of a point estimate and a confidence interval, as in traditional regression, Bayesian regression processing recovers the entire range of inferential solutions.

14.7 DISADVANTAGES OF DIFFERENT ALGORITHMS

The support vector machine (SVM) algorithm fails to produce accurate results if we work on a large data set to predict prices. It performs poorly if the data set is noisy [30].

Bayesian regression involves filtering of data and can therefore can take a long time to complete and predict. Redundancy is also low.

The ARIMA algorithm produces errors when a long-term prediction method is used [40].

The convolutional neural network (CNN) is a comparatively slower approach than other algorithms [43]. Sentiment analysis only provides a portion of the data required to accurately forecast outcomes, and many sentiment algorithms fail to distinguish between distinct sentiment targets inside the same text.

These are a few of the shortcomings of some of the approaches described above. We need to strike the ideal balance, bearing in mind the various benefits of these approaches. Some systems allow for the use of specific approaches, some of which can only be used with certain kinds of data sets. Therefore, we must ensure that the approach we select is appropriate for our system, needs, and functions (Table 14.1).

14.8 TECHNOLOGY OR EXPERIENCE?

There is always a debate around whether technology or experience is more important. Some traditional investors adhere inflexibly to orthodox training techniques and methods, whereas some budding new tech-orientated investors prefer trained bots and techniques. But the best decision is usually obtained when they go hand in hand.

Table 14.1 List of various AI, DL and ML-based strategies for crypto-currencies
price prediction

S. no.	Author	Approaches used for price prediction
1	Chen et al. [9]	Performed textual analysis to predict returns and earnings surprises.
2	Kouloumpis et al. [6]	Used standard natural language processing techniques like sentence-level and document-level scoring.
3	Alexander Pak and Patrick Paroubek et al. [7]	Created a sentiment classifier that incorporates characteristics like N-grams and POS-tags and is based on the multinomial naïve Bayes algorithm.
4	Hyunyoung Choi and Hal Varian et al. [10]	Employed seasonal autoregressive models that took Google Trends data into account.
5	Alan Dennis and Lingyao Yuan et al. [11]	Sentiment analysis by collecting valence scores on tweets.
6	Evita Stenqvist and Jacob Lonno, et al. [13]	Analyzed each tweet's sentiment and classified it as positive, neutral, or negative using VADER.
7	Connor Lamon et al. [14]	Used logistic regression to classify these tweets and then applied feature selection and classification algorithm.
8	Colianni et al. [15]	Used naïve Bayes and SVMs to identify tweets and achieve accuracy in sentiment analysis.
9	Shah et al. [16]	Used historical prices and performed Bayesian regression analysis.
10	Kimoto et al. [17]	Employed a modular neural network to forecast purchases.
11	Guresen et al. [18]	Used multilayer perceptron (MLP) neural network for its best performance.
12	Amin Azari et al. [19]	Used an ARIMA approach to predict Bitcoin prices.
13	Hari Krishnan Andi et al. [20]	Used logistic regression with LSTM machine learning model.
14	Huisu J. et al. [21]	Used Bayesian neural networks to predict Bitcoin prices.
15	L.S. Reddy et al. [22]	Used LASSO (least absolute shrinkage selection operator) AI-based algorithm.
16	Devavrat Shah et al. [23]	Tested effectiveness of Bayesian regression in predicting Bitcoin price volatility.
17	Garcia et al. [25]	Examined the relative impact of different social media sites on Bitcoin returns to see if there are any predictive links.
18	Siddhi et al. [25]	Worked on two algorithms, Bayesian regression and GLM/random forest.
19	Tian et al. [26]	Employed regression techniques and moving average values to predict Bitcoin prices, assisted by Gaussian time model.
20	Huisu Jang et al. [27]	Used a linear regression model for machine learning.

(Continued)

Table 14.1 (Continued) List of various AI, DL and ML-based strategies for
crypto-currencies price prediction

S. no.	Author	Approaches used for price prediction
21	Anshul et al. [28]	Used LSTM, a recurrent neural network algorithm, for Bitcoin price forecast.
22	Vishal et al. [29]	Employed techniques like naïve Bayes algorithm and SVM to classify the data. Using the NLTK package, discovered the polarity of tweets, which in turn was used to derive good and bad emotions.
23	Parikh and Movassate et al. [30]	Used naïve Bayes Bigram model and the maximum entropy model to classify their tweets.

Even when the market is trending downward, a good, well-informed, and well-trained trading tool can help regular investors make effective trading selections [42]. Multiple step-ahead forecasts are usually less accurate than single-step forecasts. While it is necessary to acknowledge the general support that these tools (formed by different technology stacks) provide for projecting future values, other important factors must be addressed when making decisions. Since each stock and currency has its own characteristics and may be influenced by a variety of elements that aren't always taken into account in the estimating process, a general model cannot be built from the data [38]. In spite of the advances that sometimes make us tend to rely more on technology than experience, we must also consider the accuracy of the results.

14.9 ACCURACY

Many consumers risk significant losses if the algorithms and price prediction models they rely on provide inaccurate results. The rise and fall of Bitcoin prices closely reflect the search volume index and the tweet volumes, using sentiment analysis. Price prediction accuracy is around 97.5 percent for linear regression and around 95.8 percent for decision tree [19]. In the ARIMA approach, the effect of the location and length of the test time window on the MSE achieved in price prediction was investigated. It was found that increasing the time window from 2 to 9 reduces the mean MSE of prediction error from 118000 to 16000 [19].

14.10 ETHICS

The use of these models opens up a number of ethical issues. Using Twitter data raises moral considerations about the personal space of the tweets' authors and our social responsibility to protect the privacy of those who give their data to us. Some people believe that only their followers on Twitter will see their tweets. Obtaining and interpreting these data may make something even more "public" than the author planned or understood [36].

The crypto-currencies themselves raise further ethical considerations. While many individuals favor them because they are currently uncontrolled, unregulated funds open the door for malicious or fraudulent behavior. A model to anticipate crypto-currency price movements may itself produce a false positive, making consumers more inclined to invest in crypto-currencies[48]. If the model either fails to forecast a fall in market value of the crypto-currency, or identifies it too late, as occurred in early 2018, investors can lose a lot of capital. Consequently, attempts have been made to properly elaborate what the model takes into account, as well as to accurately depict the results and the tiny sample of data on which they are based.

The algorithm should be regarded as a tool for determining what factors influence crypto-currency prices, enabling people to make more informed decisions about how they manage their money and reduce risk. Failure to recognize the risks involved or concealing these risks from others can have substantial implications, including serious financial loss, which is not something to be taken lightly [33]. Careful analysis of the data and the data points must therefore be undertaken to provide effective results.

14.11 WHY DATA ANALYSIS

Data analysis is vital since it simplifies and improves the accuracy of the data. It also helps to make data consistent by cleaning and changing the data [53]. With a more comprehensive strategy for collecting, analyzing, and presenting the data findings in a clear and understandable manner, the predictions are easier to understand [8].

14.12 CONCLUSION

There are many reasons why the precise calculation of crypto-currency prices is important. The purpose of this chapter is to provide a comprehensive study, for the benefit of new researchers, of various artificial intelligence, data learning, and machine learning techniques for crypto-currency price prediction. The chapter has assessed price prediction algorithms previously used by technologies based on AI, DL, and ML, and has listed the work done globally on crypto-currency price prediction methods, with a particular focus on significant methodologies including sentiment analysis, regression techniques, SVM, ARIMA, etc. The chapter has discussed the effect of microblogging websites like Twitter on the prices of crypto-currencies. Information about the popular crypto-currency Bitcoin and its technologies has also been provided. The extensive survey conducted for this chapter concludes that while much work has been done, more is required, and careful investigation is needed to avoid false positive results.

REFERENCES

1. Nakamoto, S.: Bitcoin: A peer-to-peer electronic cash system. In: Cryptography mailing list at https://metzdowd.com. (03 2009).
2. Ettredge, M., Gerdes, J., Karuga, G.: Using web-based search data to predict macroeconomic statistics. *Communications of the ACM*, 48(11), 87–92.
3. Miraz, M.H., Ali, M.: Applications of blockchain technology beyond cryptocurrency. *arXiv preprint arXiv:1801.03528*. COrr abs/1801.03528 (2018).
4. Hutto, C., Gilbert, E.: Vader: A parsimonious rule-based model for sentiment analysis of social media text. *8th International AAAI Conference on Weblogs and Social Media (IDWSM)*, 8(1), pp. 216–225. (2014).
5. Panger, G.T.: Emotion in social media. PhD thesis, University of California, Berkeley (2017).
6. Kouloumpis, E., Wilson, T., Moore, J. In: *Twitter Sentiment Analysis: The Good the Bad and the OMG!* AAAI Press (2011) 538–541.
7. Pak, A., Paroubek, P.: Twitter as a corpus for sentiment analysis and opinion mining. In: *LREC*. (2010).
8. O'Connor, B., Balasubramanyan, R., Routledge, B.R., Smith, N.A.: From tweets to polls: Linking text sentiment to public opinion time series. In: Proceedings of the Fourth International AAAI Conference on Weblogs and Social Media (Vol. 4, No. 1, pp. 122–129). (2010).
9. Chen, H., De, P., Hu, Y.J., Hwang, B.H.: *Customers as advisors: The role of social media in financial markets*. Working paper (2011).
10. Hyunyoung, C., Hal, V.: Predicting the present with google trends. *Economic Record* 88(s1) 2–9.
11. Sul, H., Dennis, A.R., Yuan, L.I.: Trading on twitter: The financial information content of emotion in social media. In: *2014 47th Hawaii International Conference on System Sciences* (2014) 806–815.
12. Bollen, J., Mao, H., Zeng, X.: Twitter mood predicts the stock market. *Journal of Computational Science* 2(1) (2011) 1–8.
13. Stenqvist, E., Lönnö, J.: Predicting bitcoin price fluctuation with twitter sentiment analysis. *KTH Royal Institute of Technology School of Computer Science and Communication* (2017) 3–28.
14. Lamon, C., Nielsen, E., Redondo, E.: Cryptocurrency price prediction using news and social media sentiment. Master's thesis, Stanford (2015).
15. Colianni, S., Rosales, S.M., Signorotti, M.: Algorithmic trading of cryptocurrency based on Twitter sentiment analysis *CS229 Project*, 1(5), 1–4. (2015).
16. Shah, D., Zhang, K.: Bayesian regression and bitcoin. In: *2014 53nd Annual Allerton Conference on Communication, Control, and Computing (Allerton)* (2014) 409–414.

17. Kimoto, T., Asakawa, K., Yoda, M., Takeoka, M.: Stock market prediction system with modular neural networks. In: *1990 IJCNN International Joint Conference on Neural Networks* (June 1990) 1–6. IEEE.
18. Guresen, E., Kayakutlu, G., Daim, T.U.: Using artificial neural network models in stock market index prediction. *Expert Systems with Applications* 38(8) (2011) 10389–10397.
19. Azari, A.: Bitcoin Price Prediction: An ARIMA Approach. *arXiv preprint arXiv:1904.05315.*
20. Andi, H.K.: An accurate bitcoin price prediction using logistic regression with LSTM machine learning model. *Journal of Soft Computing Paradigm (JSCP)* 03 (2021) 205–217.
21. Jang, H., Lee, J.: An empirical study on modelling and prediction of bitcoin prices with Bayesian neural networks based on blockchain information. *Digital Object Identifier* 10 (2017) 1109.
22. Reddy, L.S., Sriramya, P.: A research on bitcoin price prediction using machine learning algorithms. *International Journal of Scientific & Technology Research* 9 (2020) 1600–1604.
23. Shah, D., Zhang, K.: Bayesian regression and Bitcoin. In: Fifty-Second Annual Allerton Conference (2014).
24. Matta, M., Lunesu, I., Marchesi, M.: Bitcoin Spread Prediction Using Social and Web Search Media. In *UMAP workshops* (pp. 1–10). (2015)
25. Garcia, D., Tessone, C.J., Mavrodiev, P., Perony, N.: The digital traces of bubbles: feedback cycles between socio-economic signals in the Bitcoin economy. *Journal of the Royal Society Interface* 11 (2014) 20140623.
26. Velankar, S., Valecha, S., Maji, S.: Bitcoin price prediction using machine learning. In: 20th International Conference on Advanced Communications Technology (ICACT), pp. 144–147, IEEE.
27. Guo, T., Antulov-Fantulin, N.: Predicting short-term bitcoin price fluctuations from buy and sell orders. *Machine Learning*, arXiv:1802.04065, February 2018.
28. Jang, H., Lee, J.: An empirical study on modelling and prediction of bitcoin prices with Bayesian neural networks based on blockchain information. *IEEE Access* 6 December 4 (2017) 5427–5437.
29. Saxena, A., Sukumar, T.R.: Predicting bitcoin price using LSTM and compare its predictability with ARIMA model. *International Journal of Pure and Applied Mathematics* 119 (17) (2018) 2591–2600.
30. Vishal, A., Sonawane, S.: Sentiment analysis of twitter data: A survey of techniques. *International Journal of Computer Applications* 139 (2016) 5–15. doi;10.5120/ijca2016908625.
31. Chen, Z., Li, C., Sun, W.: Bitcoin price prediction using machine learning. *Journal of Computational and Applied Mathematics* 365 (2020) 112395.
32. Yogeshwaran, S., Kaur, M. J.: Piyush Maheshwari: Project based leaning: Predicting bitcoin prices using deep learning. In: IEEE Global Engineering Education Conference, pp. 1449–1454. IEEE. (2019).
33. Guo, T., Bifet, A., Antulov-Fantulin, N.: Bitcoin volatility forecasting with a glimpse into buy and sell orders. In: *IEEE International Conference on Data Mining (ICDM)*, pp. 989–994. IEEE. (2018).
34. Abraham, J., Higdon, D., Nelson, J., Ibarra, J.: Cryptocurrency Price Prediction Using Tweet Volumes and Sentiment Analysis. Master of Science in Data Science Southern Methodist University Dallas, Texas USA.

35. Taboada, M., Brooke, J., Tofiloski, M., Voll, K., Stede, M.: Lexicon based methods for sentiment analysis. *Computational Linguistics* 37(2) (2011) 267–307.
36. Neethu, M.S., Rajashree, R: Sentiment Analysis in Twitter using Machine Learning Techniques. In: *4th ICCCNT 2013, at Tiruchengode*, India. IEEE – 31661.
37. Nofer, M: The value of social media for predicting stock returns: Preconditions, instruments and performance analysis. Springer, 2015, doi:10.1007/978-3-658-09508-6.
38. Siddhi V., Sakshi V., Shreya M.: Bitcoin price prediction using machine learning. In: *International Conference on Advanced Communications Technology (ICACT)* 144–147. IEEE.
39. Andrade de Oliveira, F., Enrique Zárate, L., de Azevedo Reis, M.; Neri Nobre, C.: The use of artificial neural networks in the analysis and prediction of stock prices. In: IEEE International Conference on Systems, Man, and Cybernetics, 2011, pp. 2151–2155. IEEE.
40. Daniela, M., Butoi, A.: Data mining on Romanian stock market using neural networks for price prediction. *Informatica Economica* 17, 2013.
41. Mahale, R.R.: *Bitcoin Price Prediction*, NHCE, 2018.
42. Armano, G., Marchesi, M., Murru, A.: A hybrid genetic-neural architecture for stock indexes forecasting. *Information Sciences* 170 (2005) 3–33.
43. Liaoa, Z.: Forecasting model of global stock index by stochastic time effective neural network. *Expert Systems with Applications* 37(1) (2010) 834–841.
44. Change, P.-C.: A neural network with a case based dynamic window for stock trading prediction. *Expert Systems with Applications* 36(3) (2009) 6889–6898.
45. Zhanga, Y.: Stock market prediction of S&P 500 via combination of improved BCO approach and BP neural network. *Expert Systems with Applications* 36(5) (2009) 88498854.
46. Sánchez, E., Olivas, J.A., Francisco, P.: Romero data analytics for the cryptocurrencies behaviour. In *Cloud Computing and Big Data: 7th Conference, JCC&BD 2019, La Plata, Buenos Aires, Argentina, June 24–28, 2019, Revised Selected Papers 7* (pp. 86–97). Springer International Publishing.
47. Wilson-Nunn, D., Zenil, H.: On the complexity and behaviour of cryptocurrencies compared to other markets. arXiv:1411.1924 [q-fin.ST], 2014
48. Kim, Y.B., Kim, J.G., Kim, W., Im, J.H., Kim, T.H., Kang, S.J.: Predicting fluctuations in cryptocurrency transactions based on user comments and replies. *PLoS One* 11(8) (2016) e0161197.
49. Aslanidis, N., Bariviera, A.F., Martinez-Ibañez, O.: An analysis of cryptocurrencies conditional cross correlations. *Finance Research Letters* 31 (2019) 130–137.
50. Rathan, K., Sai, S.V., Manikanta, T.S.: Crypto-currency price prediction using decision tree and regression techniques. In *Third International Conference on Trends in Electronics and Informatics (ICOEI 2019) IEEE Xplore Part Number: CFP19J32-ART*; ISBN: 978-1-5386-9439-8
51. Neil, G., Halaburda, H.: Can we predict the winner in a market with network effects? Competition in cryptocurrency market. *Games* 7(3) (2016) 16.
52. https://www.analyticsfordecisions.com/importance-of-data-analysis-inresearch/#:~:text=Data%20analysis%20is%20important%20in,them%20derive%20insights%20from%20it

53. https://paraphrase.projecttopics.org/importance-of-data-analysis-in-a-researchpaper.html
54. Vockathaler, B.: The bitcoin boom: An in depth analysis of the price of bitcoins. Thesis, University Of Ottawa, Ontario, Canada, June 2017.
55. Madan, I., Saluja, S., Zhao, A.: Automated Bitcoin Trading via Machine Learning Algorithms. Department of Computer Science, Stanford University, Stanford, 2015.

Chapter 15

Interactive remote control interface design for a cyber-physical system

Wei-Chih Chen
National Yang Ming Chiao Tung University, Hsinchu, Taiwan

Hwang-Cheng Wang
National Ilan University, Yilan, Taiwan

Abhishek Sharma
The LNM Institute of Information Technology, Jaipur, India

CONTENTS

15.1 INTRODUCTION

Ever since the confirmation of COVID-19 in December 2019, the disease has become one of the most contagious pandemics in human history, with a high rate of mortality and a rapid spread across the globe. COVID-19 is mainly transmitted through personal contact and airborne transmission. Unlike in the past, the virus can survive outside the host for a few hours. As a result, indirect contact transmission may also occur. The best way to avoid infection is thus to stay away from public places and reduce personal interaction.

Despite the pandemic, most industries needed to continue operations. This is especially true for companies with vital facilities that must be maintained in good condition. In addition, high temperatures and pressures may make certain spots inaccessible to human beings. Situations may also arise in a company or factory requiring urgent resolution. Almost all the standard remote operation tools available on the market only support a virtual interface specifically tailored to the system and are insufficiently flexible.

DOI: 10.1201/9781003415466-15

The COVID-19 pandemic has further exacerbated the shortage of competent human workers.

In this study, we devised a human–machine interface that allows interaction remotely. Bearing in mind the diversity of needs and tasks in different situations, we designed the mechanism, adjusted the parameters, and tested it in a simulation environment before deployment in a real environment. The prototype system comprises a camera and a robotic arm. By providing real-time video and an intuitive control interface through the high-speed network, the camera serves as the operator's eyes, and the robotic arm replaces the operator's hands to provide assistance or complete simple tasks from a distance to achieve remote operation of real-world objects. We built a Gazebo simulation environment based on ROS and integrated the robotic arm, YOLOv3, and adaptive tracking to design an interface for remote control and observation by users. The virtual environment was used for experimentation and testing, and the control parameters and models can be deployed in real scenarios.

The chapter is structured as follows. Section 15.2 provides a brief review of related works. The methodology employed is described in Section 15.3. Results and analysis are presented in Section 15.4. Section 15.5 discusses the implications of deep learning in the development of CPS. Finally, Section 15.6 concludes the chapter and points to future work.

15.2 RELATED WORKS

Image processing and its applications have played essential roles in the development of many technologies. Since the 1970s, digital image processing has gained prominence because of the many advantages it offers. In recent years, the outstanding performance of digital image processors has strengthened their position in daily life and industry. It is easy to exploit these advantages for object tracking.

In an intelligent video surveillance system, object tracking is a significant and crucial function. Many algorithms, such as the Kalman filter, particle filter, and mean shift, were created specifically for tracking. However, since all these algorithms have their limitations, it is not effective to use only one of them. Iswanto et al. [1] devised an algorithm that incorporated them all and combined the color histogram and texture characteristics to improve accuracy. In the current literature, techniques and approaches for object tracking, which is also common in intelligent video surveillance, are broadly divided into two categories: hand-crafted video features [2, 3], and representation learning based on deep learning architectures [4, 5].

Tsai [6] used binarization and mid-pass filtering to achieve the separation of target and background and then employed image subtraction and moving edge detection to find the position of a target in an image. Afterwards, the difference in the position of the object in consecutive frames is used to determine the rotation parameters of a two-axis machine. Although this approach is feasible, it can only operate on a simple background. When a complex texture or a strong light source appears in the background, it is prone to misjudgment. Xie [7] applied the video tracking technique to childcare. The system generates alerts via cell phones when infants or toddlers move out of a safe range. Similar methods are used in [6, 7] for object segmentation. However, they both suffer the problem of poor performance under adverse conditions. In this study, we found that neural networks show better fault tolerance, scalability, and flexibility than conventional methods.

Redmon and Farhadi proposed YOLOv3 in 2018 [8]. Although the network size is larger than YOLOv2, accuracy and precision are significantly improved, and the problem of poor performance in detecting small objects is overcome. To improve efficiency, we use the Robot Operating System (ROS) with the Gazebo simulator for model training, testing, and simulation of mechanical motion [9]. Then the trained models and algorithms are deployed on real-world machines.

15.3 RESEARCH METHODOLOGY

This study uses YOLOv3 for object recognition and tracking. Since object recognition results are vulnerable to interference from the light source, shaking, target texture, or camera quality, we first built a simulator and carried out the entire study in the simulation environment. The simulation included sampling, training, and testing (Figure 15.1). The construction of the simulation environment was based on Gazebo, and a controller running the Ubuntu operating system managed the system in conjunction with ROS (Figure 15.2). The camera can be controlled manually or by using the YOLOv3 auto-tracking mode.

Figure 15.1 System flowchart.

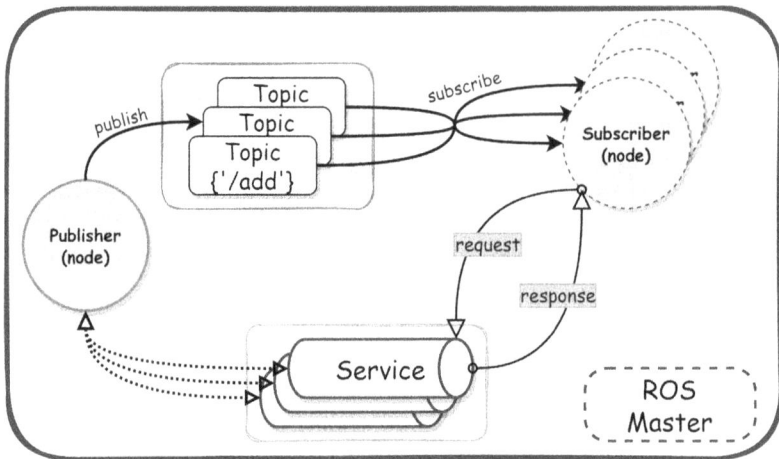

Figure 15.2 ROS I architecture.

The robot operating system (ROS) is a set of software libraries and tools that help developers build robot applications. It is based on a distributed system architecture, where the various components of a robot application are distributed across multiple processes and communicate with each other using a publish/subscribe messaging model. The ROS master acts as a central directory for the system, and nodes are the basic building blocks that communicate with each other using messages. ROS 1 also includes additional tools and libraries for common functionality, such as package management, message passing, and visualization. Overall, the ROS 1 architecture is designed to be flexible and modular, making it easy for developers to build and scale robot applications.

ROS integrates many features such as hardware abstraction, driver management, messaging between programs, execution of shared functions, distribution suite management, and a library of tools to extract, build, write, and run on a hardware platform. ROS supports multiple programming languages such as C++, Python, Java, etc. The system uses the Ubuntu architecture, and the main programming languages are C/C++ and Python, which are very suitable for use with ROS for the robotic arm and video streaming control.

Gazebo provides software-in-the-loop (SIL) simulation, which allows developers to simulate in a non-real-time modeling environment. Actual physical parameters are introduced into the simulator, and test results from the simulator can be treated as real-world products.

YOLOv3 is much more complex than its predecessor. However, the improvement in speed and accuracy is also significant [8]. Moreover, by altering the structure of the model, it is possible to deploy it on embedded systems with constrained computing resources.

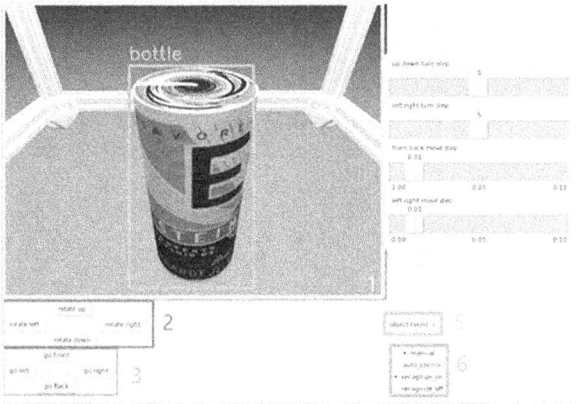

Figure 15.3 Remote human–machine interface.

The graphical user interface (GUI) is designed to provide the user with a real-time view of the position of the end-effector in the simulation environment and the scene captured by the camera. This allows the user to intuitively control the attitude and position of the end-effector and select between the "automatic object tracking" and "manual control" functions via the interface.

The pink box (1) in Figure 15.3 is the first-person view image from the camera on the end-effector. The buttons in the red box (2) are used to control the rotation direction of the end-effector. The blue box (3) is used to control the end-effector to shift forward, backward, left, or right on the plane coordinates. The slider in the yellow box (4) is used to adjust the distance of movement and angle of rotation during manual control. The number shown in the orange box (5) is the number of objects recognized on the screen. Finally, the green box (6) allows one to select whether or not to enable automatic object tracking and whether or not to frame the objects and display the tag names on the screen (1).

In this work, we designed an algorithm (Figure 15.4) for the adaptive tracking of the target object. The bounding box information of the object recognition result by the Yolov3 model is used to determine the location of the target object. Then, the end-effector is moved adaptively to keep the target object at the center of the returned image. The procedures are described in the following.

First, the model recognizes the target object and outputs the coordinates of the four corners of the bounding box.

$$B_p = \begin{bmatrix} x_1 \\ x_2 \\ y_1 \\ y_2 \end{bmatrix} \tag{15.1}$$

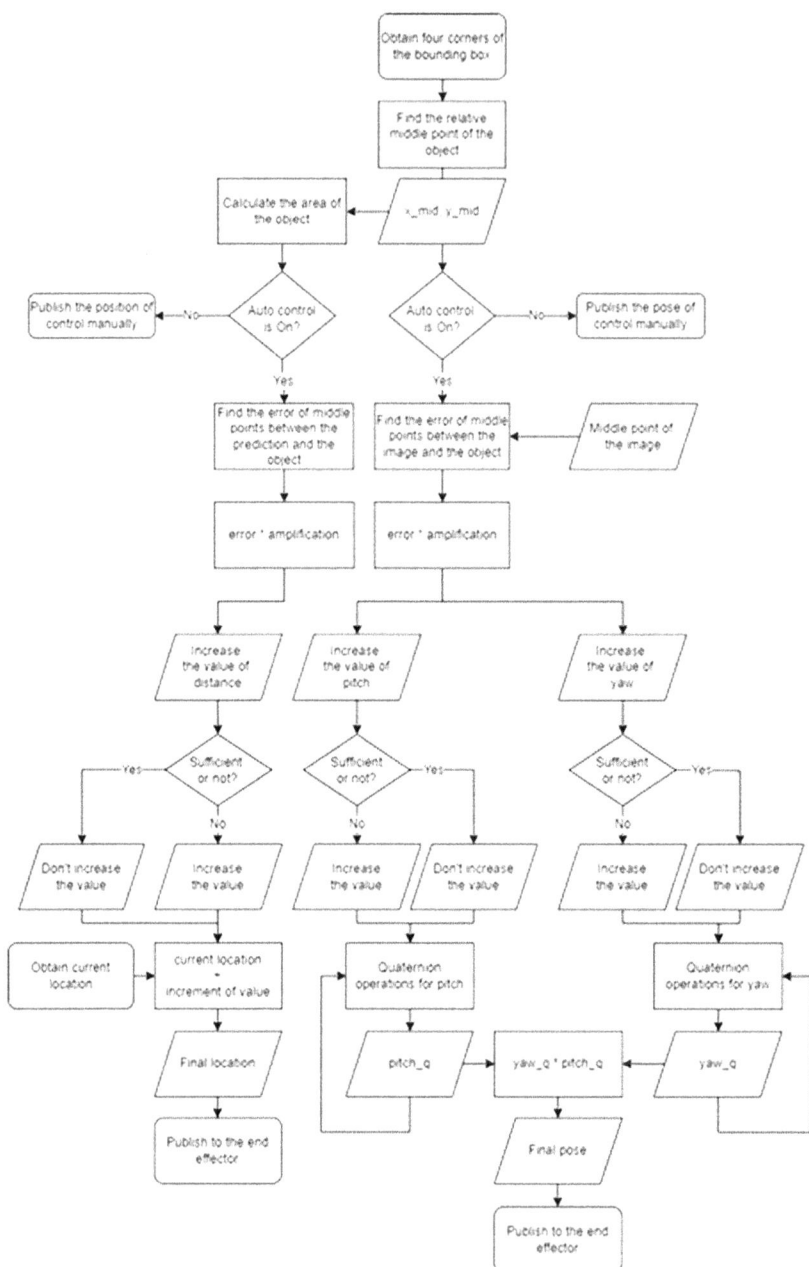

Figure 15.4 Strategic control flowchart.

After the center of the bounding box (x_{mid}, y_{mid}) is determined, the distance (x_e, y_e) from the center of the screen (x_r, y_r) is calculated. Then the distance is multiplied by the factor k_1 to get the angle $(\Delta\varphi, \Delta\psi)$ by which the end-effector should be rotated. Adding the rotation angle to the original angle $(\varphi(k), \psi(k))$ gives the expected angle $(\varphi(k + 1), \psi(k + 1))$ of the end-effector at the next time instance, which is then published to the end-effector in the simulation environment.

The area of the bounding box B_a is used to decide whether the end-effector should increase or decrease the distance to the target object. By calculating the area obtained and subtracting it from the target area T_a (the area of the object that should be reached), the area difference E_a is obtained, which is then multiplied by the factor k_2 to get the moving distance Δz of the end-effector. The moving distance plus the original position $z(k)$ gives the expected position of the end-effector at the next time unit $z(k + 1)$, which can then be published to the end-effector in the simulation environment.

The equations pertaining to the above procedures are given below:

$$\begin{bmatrix} x_{mid} \\ y_{mid} \end{bmatrix} = \begin{bmatrix} \left(B_p[1] + B_p[2]\right)/2 \\ \left(B_p[3] + B_p[4]\right)/2 \end{bmatrix} \tag{15.2}$$

$$x_e = x_r - x_{mid} \tag{15.3}$$

$$\Delta\varphi = k_1 \times x_e \tag{15.4}$$

$$\varphi(k+1) = \varphi(k) + \Delta\varphi \tag{15.5}$$

$$y_e = y_r - y_{mid} \tag{15.6}$$

$$\Delta\psi = k_1 \times y_e \tag{15.7}$$

$$\psi(k+1) = \psi(k) + \Delta\psi \tag{15.8}$$

$$B_a = \left(B_p[1] - B_p[2]\right) \times \left(B_p[3] - B_p[4]\right) \tag{15.9}$$

$$E_a = T_a - B_a \tag{15.10}$$

$$\Delta z = k_2 \times E_a \tag{15.11}$$

$$z(k+1) = z(k) + \Delta z \tag{15.12}$$

15.4 RESULTS AND DISCUSSION

In order to improve efficiency and reduce the design cost, this study built a scalable cube frame in the Gazebo simulator (Figure 15.5). A camera was also mounted to provide a first-person perspective of the view (Figure 15.6). ROS would send the camera image to YOLOv3 for recognition and then to the control interface for user observation and operation. Whether the camera changes its position through the control interface or the target object is moved through the Gazebo simulator interface, a series of coordinate conversions and processing automatically adjust the position of the end-effector so that the target object remains at the center of the screen.

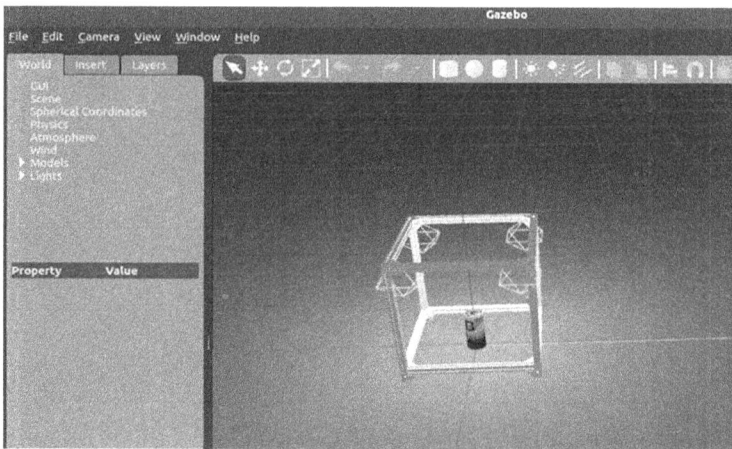

Figure 15.5 The cube frame in Gazebo, in this case with a beer can as the target object.

Figure 15.6 End-effector and a first-person view of the screen.

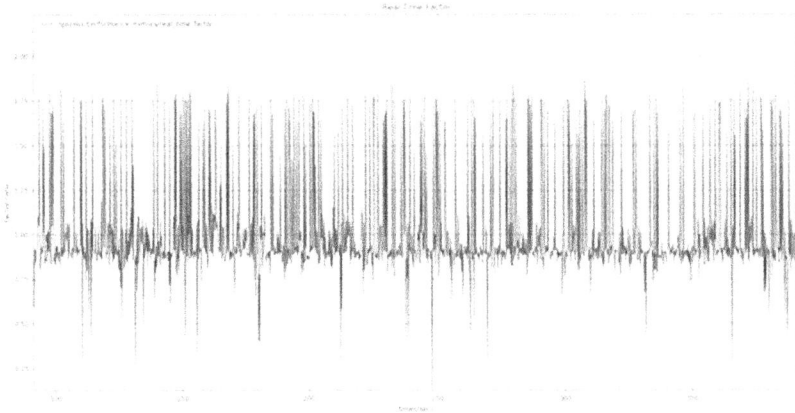

Figure 15.7 Real-time factor of the Gazebo simulation environment.

The execution of the system starts with the activation of roscore and Gazebo_simulator, and then Darknet_YOLO. It takes about one minute for the real-time factor (RTF) to stabilize. RTF indicates the performance of the simulation and is used as an efficiency benchmark [10]. Figure 15.7 shows the RTF of the Gazebo simulation. An RTF of 1 means synchronization with physical time, and a larger value indicates that the simulation is more efficient. As shown in the figure, the actual operation lasts about six minutes, and the RTF is greater than 1 most of the time, indicating high efficiency. Note that the RTF is affected by the hardware, the number of objects in the simulation environment, and the complexity of the task.

Figure 15.8 shows the change in the position of the target object and the camera with time under the auto-tracking mode of operation. In the figure,

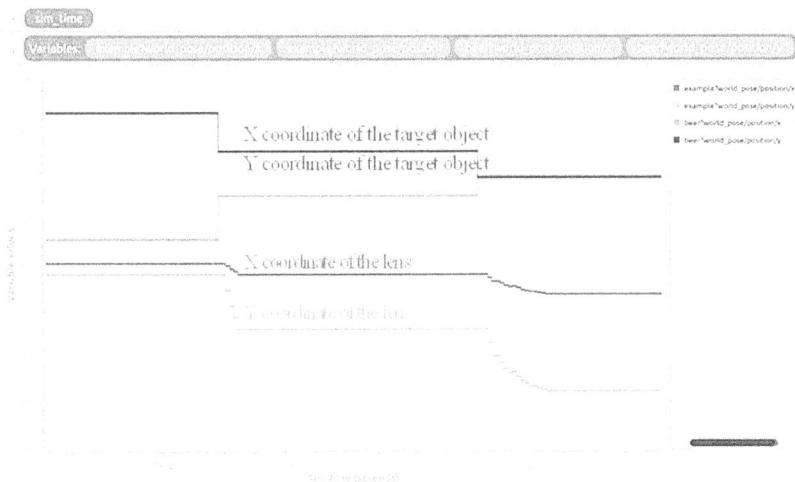

Figure 15.8 Position of the lens and the target object change in auto-tracking mode.

the horizontal axis is the time, and the vertical axis is the X and Y coordinate values of the absolute position in the environment. Observe that after the target object moves, the camera quickly makes a corresponding position change in auto-tracking mode.

Three desirable features of the implementation are summarized below.

- Different models or parameters can be chosen to identify and track different types of target objects.
- The cube frame can be extended or contracted at will and can be resized according to the task and demand.
- The cost of developing, modifying, and conducting experiments is much lower than that of traditional processing and manufacturing methods. In addition, the simulation environment can provide a large number of training materials and backgrounds [11], thereby reducing the cost of data collection.

15.5 DEEP LEARNING APPLICATIONS IN CPS

Due to advances in technology areas such as Industry 4.0, sustainable smart societies (S3), digital contracts, Metaverse, etc., intelligence applications play an essential role. Not only do they fetch data from previously inaccessible places, but they also contribute to the system's overall performance. Intelligence can be applied to CPS in many areas such as automation, diagnosis, and decision-making with and without human intervention.

Deep learning techniques are significantly used in such environments. In [12], IoT infrastructure is proposed with a distributed communication environment using multiple communication technologies for CPS. Distributed communication environments are valuable in large-scale systems. The work presented by [13] shows the comparative analysis of deep learning approaches and the concept of secure deep learning methods to prevent data augmentation and malware attacks. Due to the design requirement and interconnectivity, security is an essential requirement of CPS systems. Secure deep learning mechanisms are an important tool for making systems less vulnerable to attacks. CPS systems generate data in various forms, which will be further used for analysis and decision-making. Reference [14] presents anomaly detection using deep learning. Identifying outliers can be very helpful for pointing out alerts and irregularities in the overall work environment of CPS.

15.6 CONCLUSION AND FUTURE WORK

This research integrates the control and image recognition mechanism into the Gazebo environment. The conditions in the simulation environment are similar to those in reality, and the ROS subscription and publication mechanism makes the whole system scalable, easy to modify, and well integrated. In the future, we will explore the possibility of directly connecting the system to real robot arms to achieve simultaneous simulation and control of a cyber-physical system. Thanks to rapidly advancing digital technology, hardware costs continue to fall, leading to the popularity of smartphones and other mobile devices, which can facilitate such remote operations. The use of smartphones for the manipulation of cyber-physical systems will also be investigated.

ACKNOWLEDGMENT

This research was funded by the Ministry of Science and Technology, Taiwan, ROC, contract number MOST 110-2813-C-197-015-E.

REFERENCES

[1] Iswanto, I. A., Choa, T. W., & Li, B. (2019). Object tracking based on meanshift and particle-Kalman filter algorithm with multi features. *Procedia Computer Science*, vol. 157, pp. 521–529.

[2] Ojha, S., & Sakhare, S. (2015). Image processing techniques for object tracking in video surveillance-A survey. *In Proceedings of the International Conference on Pervasive Computing (ICPC)*, Pune, India, pp. 1–6.

[3] Gowsikhaa, D., Abirami, S., & Baskaran, R. (2014). Automated human behavior analysis from surveillance videos: a survey. *Artificial Intelligence Review*, vol. 42(4), pp. 747–765.

[4] Hasan, M., Choi, J., Neumann, J., Roy-Chowdhury, A. K., & Davis, L. S. (2016). Learning temporal regularity in video sequences. In *Proceedings of the IEEE Conference on Computer Vision and Pattern Recognition*, Las Vegas, USA, pp. 733–742.

[5] Nawaratne, R., Bandaragoda, T., Adikari, A., Alahakoon, D., De Silva, D., & Yu, X. (2017). Incremental knowledge acquisition and self-learning for autonomous video surveillance. *In Proceedings of the 43rd Annual Conference of the IEEE Industrial Electronics Society*, Beijing, China, pp. 4790–4795.

[6] Tsai, B. C. (2002). A Study of a Monitoring and Control System Using Image Tracking Methods, Master's Thesis, Institute of Mechanical Engineering, Chung Yuan Christian University, Taoyuan, Taiwan.

[7] Hsieh, L. C. (2014). The Study of Image Track and Surveillance Monitor System, Master's Thesis, Department of Electronic Engineering, National Kaohsiung University of Applied Science and Technology, Kaohsiung, Taiwan.

[8] Redmon, J., & Farhadi, A. (2018). Yolov3: An incremental improvement. preprint arXiv:1804.02767.

[9] Bu, Q., Wan, F., Xie, Z., Ren, Q., Zhang, J., & Liu, S. (2015). General simulation platform for vision based UAV testing. *In Proceedings of IEEE International Conference on Information and Automation*, Lijiang, China, pp. 2512–2516, DOI: 10.1109/ICInfA.2015.7279708.

[10] Abbyasov, B., Lavrenov, R., Zakiev, A., Yakovlev, K., Svinin, M., & Magid, E. (2020). Automatic tool for gazebo world construction: From a grayscale image to a 3d solid model. *In Proceedings of the IEEE International Conference on Robotics and Automation (ICRA) Paris*, France, pp. 7226–7232.

[11] Borrego, J., Figueiredo, R., Dehban, A., Moreno, P., Bernardino, A., & Santos-Victor, J. (2018). A generic visual perception domain randomisation framework for Gazebo. *In Proceedings of the IEEE International Conference on Autonomous Robot Systems and Competitions (ICARSC), Torres Vedras*, Portugal, pp. 237–242, DOI: 10.1109/ICARSC.2018.

[12] Singh, S. K., Jeong, Y.-S., & Park, J. H. (2020). A deep learning-based IoT-oriented infrastructure for secure smart city, *Sustainable Cities and Society*, vol. 60, p. 102252.

[13] Hatcher, W. G., & Yu, W. (2018). A survey of deep learning: platforms, applications and emerging research trends, *IEEE Access*, vol. 6, pp. 24411–24432.

[14] Luo, Y., Xiao, Y., Cheng, L., Peng, G., & Yao, D. (2022). Deep learning-based anomaly detection in cyber-physical systems, *ACM Computing Surveys*, vol. 54, no. 5, pp. 1–36.

Chapter 16

Collateral-based system for lending and renting of NFTs

Tarun Singh, Meet Kumar Jain, Vibhav Garg,
Vinay Kumar Sharma, Punit Agarwal, Vikas Bajpai
and Abhishek Sharma

The LNM Institute of Information Technology, Jaipur, India

CONTENTS

DOI: 10.1201/9781003415466-16

16.1 BACKGROUND

16.1.1 Blockchain [1]

Before looking at what blockchain is, consider the following scenario.

Currently, your data is stored at a particular place in a company but are you satisfied that your data is secure there? As all the information is in one place, it becomes a massive target for hackers. Also, do you trust the company to maintain the data for you? What if some day the company betrays you and sells your data?

So how does blockchain address this issue? We first need to understand the term "centralization." Centralization refers to the control of activity under a single authority. All trust proofs are stored and signed by single entities in closed databases around the globe.

Issues with centralization are the following:

- *Security and privacy*. Organizations have been victims of hacking attacks or data leaking.
- *Traceability*. Having a private database makes it hard for companies, users, and the government to keep track of the validity of the data and share data consistently.
- *High costs*. The charges made by intermediaries.
- *Speed*: Centralized institutions are slow because intermediaries must verify data validity through cumbersome processes.

16.1.2 How do we solve the problem of centralization?

The answer is decentralization, i.e., removing the intermediaries and creating trust via tamper-proof, irreversible, automated, and distributed technology-driven processes. This results in what is known as blockchain. The blockchain is a distributed network of nodes that share a public database containing encrypted data and information about that data's transactions (Figure 16.1).

Notable features of blockchain are that it is:

- Secure
- Decentralized
- Transparent
- Efficient
- Fast

Interest in Blockchain over the years

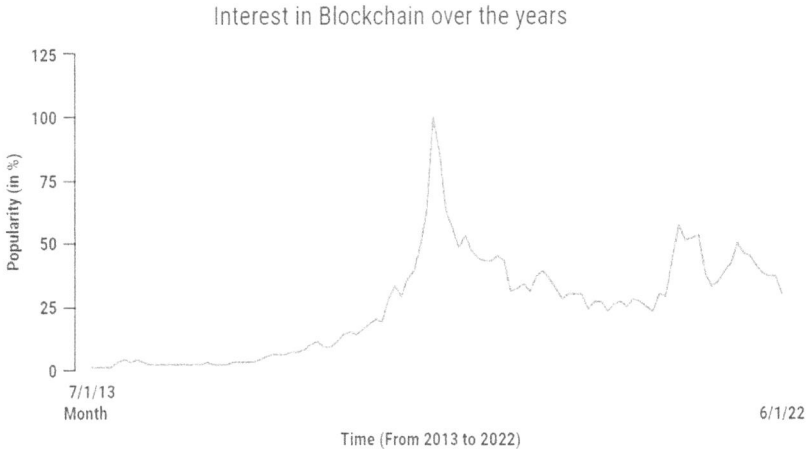

Figure 16.1 Interest in blockchain over time. (Source: Google Trends.)

16.1.3 Blockchain applications

Blockchain is a distributed and decentralized database model that has great potential to unravel a number of privacy and trust problems. Currently, we have only seen blockchain used in cryptocurrencies like bitcoin or ethereum, but blockchain also has many uses in finance, supply chains, etc.

- *Blockchain and IoT Integration.* Blockchain technology is highly integrable with the Internet of Things (IoT) as transaction history between two users can be created.
- *Blockchain in vaccine manufacture and tracking.* Blockchain technology has many potential use cases in vaccine tracking and distribution. In a potential future scenario of the rapid creation of fake vaccines and medicines, blockchain technology to ensure the authenticity of drugs and vaccines can come into play. It can also help with efficient medicines supply chains (Figure 16.2).
- *NFTs become the Next Big Thing.* Non-fungible tokens, or NFTs, are one of the hottest topics in the blockchain industry. Massive sums are invested in NFTs: in 2021, an NFT was sold for $69 million.

16.1.4 Non-fungible tokens (NFTs)

The NFT is a digital asset representing real-world assets like art, music, in-game items, and videos [2]. An NFT is created or "minted" from digital assets that represent tangible and intangible assets, including:

- Works of art
- GIFs
- Videos or sports clips
- Collectibles

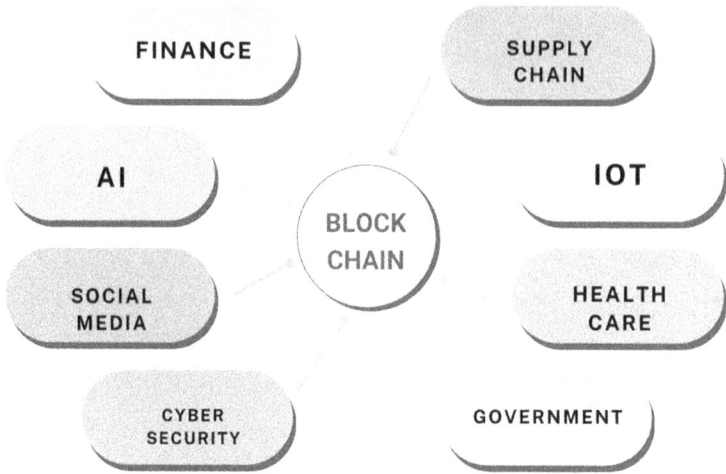

Figure 16.2 Applications of blockchain technology.

- Virtual avatars and video game skins
- Designer sneakers
- Music

NFTs are currently taking the digital art and collectibles world by storm. Digital artists and content creators are seeing their lives change due to significant sales to a new crypto-audience. Many big names have spotted a brand-new opportunity to connect with fans and are also joining this NFT train. But digital art is just one way to use NFTs. They can represent ownership of any unique asset, a deed for an item within the digital or physical realm. These things are not interchangeable with other items because they have unique properties.

16.1.5 Trends in the NFT market

The NFT boom began in 2021, with pfp as nft format, then slowly shifted towards art and picture NFTs. We live in the age of in-game items, music NFTs, NFT tickets, etc. Now music artists prefer to launch an NFT for their new album rather than posting the music on streaming apps (Figures 16.3 and 16.4).

16.2 COLLATERAL-BASED SYSTEM IN TRADITIONAL BANKS [6]

Two kinds of security are commonly used in bank loans, personal security and collateral. Banks use different methods to take collateral from customers. The term "personal security" is used for loans for personal items like houses, cars, etc. Collateral is also required when taking out a business loan.

INTEREST IN NFTS

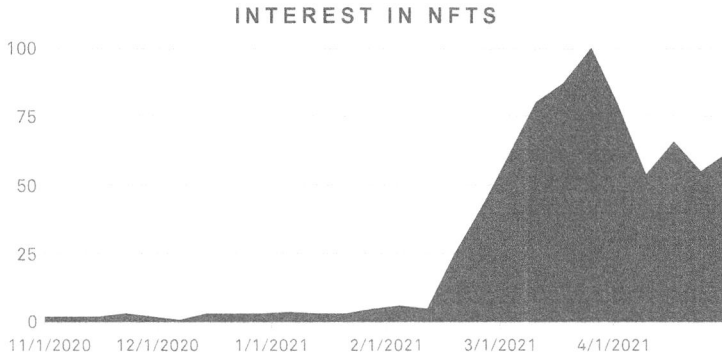

Figure 16.3 Interest in NFTs over the years. (Source: Google Trends.)

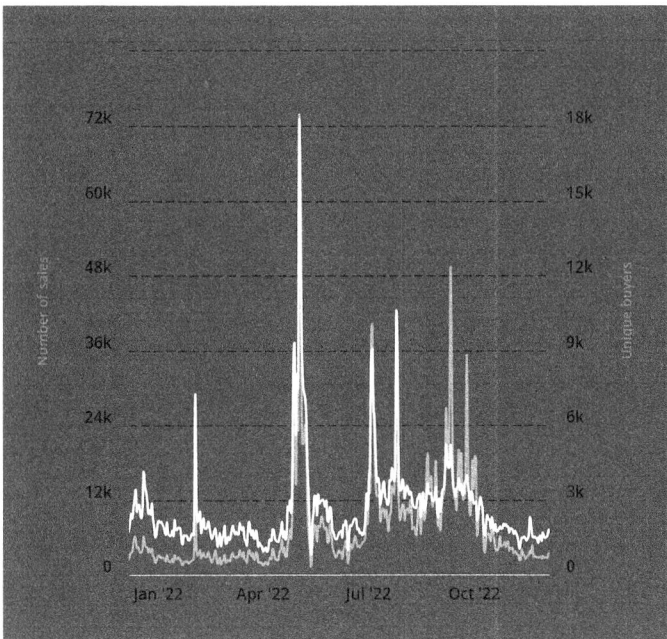

Figure 16.4 Number of sales of NFTs in 2022.

Types of Collateral

- *Real estate.* The most common type of collateral borrowers use in the real world is real estate, such as a piece of land or a home. Such properties come with a very high value and low depreciation.
- *Cash secured loan.* Cash is another of the most widely used types of collateral due to how easily it works. A person can easily take a loan from the bank where they maintain an account, and if they default, the bank can liquidate their accounts to recover the borrowed money.

- *Inventory financing.* This type involves an inventory that serves as the collateral for a loan.
- *Invoice collateral.* Invoices are one of the types of collateral used by small businesses, where invoices to company customers that are still outstanding – unpaid – are used as collateral.
- *Blanket liens.* This involves using a lien, a kind of legal claim that allows a lender to dispose of the assets of a business that has defaulted on a loan.

16.3 OVERVIEW

16.3.1 Problem statement

The major problem with NFTs is their skyrocketing prices, which makes them difficult to follow, especially for the average person. And the only appreciation the owner receives on the NFT is when it resells at a higher price than he paid for it (Figure 16.5). We have tried to develop a solution that solves both these issues.

16.3.2 Proposed solution

Owners who wish to rent out their NFTs can list them on the platform by specifying the collateral amount, the daily rent charge, and the maximum number of days one can rent that NFT. The renters who come across a particular NFT of interest can then initiate the NFT renting process by specifying the number of days they wish to borrow the NFT. After depositing the collateral and renting charge, the collateral and renting amount is deposited

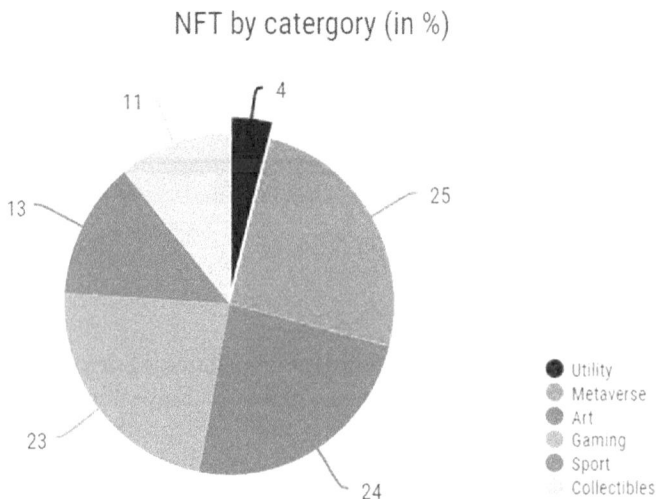

Figure 16.5 Distridrawn by ownbution of NFTs in different categories. (Source: nonfungible.com.)

into a smart contract with predefined terms. The ownership of the NFT is transferred to the renter and the renting charge to the lender. The renter then returns the NFT after the renting period and collects the collateral.

16.4 NFT LIFECYCLE

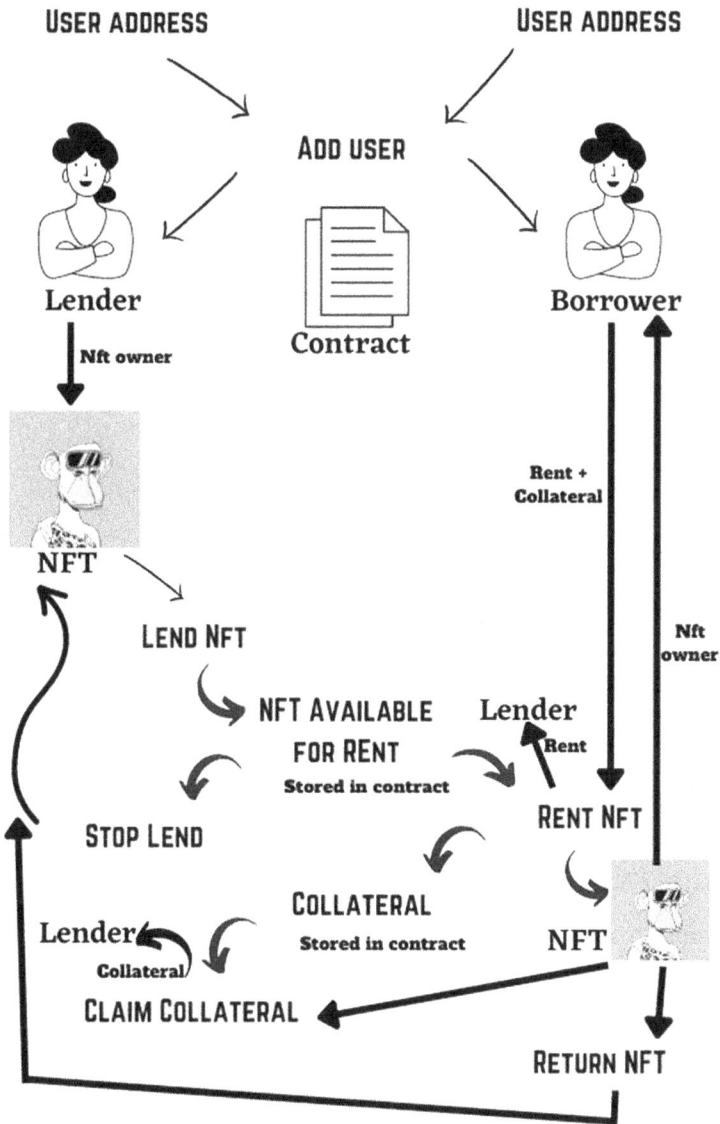

Figure 16.6 Lifecycle of NFT in our project.

16.5 CONTRACT OVERVIEW

16.5.1 Getters

Getters and setters play an important role in retrieving and updating the value of a variable outside the encapsulating class. For each instance variable, a getter method returns its value.

16.5.1.1 Get NFT details

This gives the details of an NFT, such as NFT OWNER, NFT COLLECTION ADDRESS, NFT token id, etc. when given the NFT key.

> **Getter 1**: Get NFT details
>
> **Input**: Unique NFT key (_nftKey)
> **Output**: NFT details:
> * NFT owner (_nftOwner)
> * NFT collection address (_nftAddress)
> * NFT token ID in the collection (_nftId)
> * Other NFT attributes (_nftName, _nftImageURL)

16.5.1.2 Get NFT list available for rent

This gives the list of NFTs available for rent in our dApp.

> **Getter 2**: Get NFT list available for rent
>
> **Output**: availableToRentNftList

16.5.1.3 Get lend NFT details

This gives the details of the NFT lent, like lender address, borrower address which is initially null or address(0) when there is no borrower, time for which NFT is being lent, the daily price of the NFT, and the collateral requested when given the NFT key.

> **Getter 3**: Get lent NFT details
>
> **Input**: Unique NFT key (_nftKey)
> **Requirements**: availableToRentNftList.contains (_nftKey)
> **Output**:
> * Lender address (_lenderAddress)
> * Borrower address (_borrowerAddress)
> * Time for which NFT is being lent (_dueDate)
> * Daily rent price of NFT (_dailyRent)
> * Collateral required (_collateral)

16.5.1.4 Get rent NFT details

This gives the details of the NFT rented, such as lender address, borrower address, number of days for which NFT is rented, and the start time of renting NFT when given the NFT key.

Getter 4: Get rent NFT details

Input: Unique NFT key (_nftKey)
Requirements:
- nftList.contains (_nftKey)
- not (availableToRentNftList.contains (_nftKey))
Output:
- Lender address (_lenderAddress)
- Borrower address (_borrowerAddress)
- Number of days for which NFT rented (_numberOfDays)
- Start time of NFT renting (_rentalStartTime)

16.5.1.5 Get user lend NFT details

This gives the details of the NFT lent by a particular lender/user given its address and the corresponding NFT key. Details of the NFT lent include lender address, borrower address which is initially null or address(0) when there is no borrower, time for which NFT is being lent, the daily price of the NFT, and the collateral requested.

Getter 5: Get user lent NFT details

Input:
- Unique NFT key (_nftKey)
- User address (_userAddress)
Requirements:
- _userAddress != address(0)
- User exists (_userAddress)
- _userAddress.lentList.contains (_nftKey)
Output:
- Lender address (_lenderAddress)
- Borrower address (_borrowerAddress)
- Time for which NFT is being lent(_dueDate)
- Daily rent price of NFT (_dailyRent)
- Collateral required (_collateral)

16.5.1.6 Get user rent NFT details

This gives the details of the NFT rented by a particular borrower/user given its address and the corresponding NFT key. Details of the NFT rented include lender address, borrower address, number of days for which NFT is rented, and the start time of renting NFT.

Getter 6: Get user rent NFT details

Input:
- Unique NFT key (_nftKey)
- User address (_userAddress)

Requirements:
- _userAddress != address(0)
- User exists (_userAddress)
- _userAddress.rentList.contains (_nftKey)

Output:
- Lender address (_lenderAddress)
- Borrower address (_borrowerAddress)
- Number of days for which NFT rented (_numberOfDays)
- Start time of NFT renting (_rentalStartTime)

16.5.2 Algorithms

16.5.2.1 Add user

This functionality adds a new user to the dApp.

Algorithm 1: Add User

Input: User address to be added (_userAddress)
Requirements:
- _userAddress != address(0)
- msg.sender ==_userAddress

Procedure:
- if not(user already exists) **then**
    newUser = createNewUser()
    usersList.add(newUser)
  end

16.5.2.2 Lend NFT

This functionality is used when a user (lender) wants to put their NFT on lend to earn rent on the NFT. The user gives details like due date, i.e., time for which they want to put their NFT on lend, daily rent price, and collateral amount requested.

We create an instance of the ERC721 standard, which is the standard token standard for NFTs, and call the transfer function of that NFT from lender address to contract address. All the NFTs lent stay in the contract unless someone rents it.

Algorithm 2: Lend NFT

Input:
- Unique NFT key (_nftKey)
- NFT owner (_nftOwner)
- NFT collection address (_nftAddress)
- NFT token ID in the collection (_nftId)
- Other NFT attributes (_nftName, _nftImageURL)
- Lender address (_lenderAddress)
- Time for which NFT is being lent (_dueDate)
- Daily rent price of NFT (_dailyRent)
- Collateral required (_collateral)

Requirements:
- msg.sender == _lenderAddress
- User exists (_lenderAddress)
- _nftKey should not be already lent
- _nftOwner == _lenderAddress
- _dueDate > current time
- Contract address must be approved for Nft transfer

Procedure:
- ERC721 nftCollection = ERC721 (_nftAddress)
- nftCollection.safeTransferFrom   (_lenderAddress,   contract address, _nftId)
- nftList.add (_nftKey)
- _lenderAddress.lentList.add (_nftKey)
- availableToRentNftList.add (_nftKey)

16.5.2.3 Rent NFT

This functionality is used when a user (borrower) wants to rent an NFT. The user gives the number of days for which they want to rent the NFT as input and pays the requested amount of ether, i.e., NFT rent price + collateral. The number of days for which the NFT can be rented cannot be greater than the number of days the NFT is available to rent.

We create an instance of the ERC721 standard and call the transfer function of that NFT from the contract address to the borrower address. The collateral paid by the borrower now resides in the contract.

Algorithm 3: Rent NFT

Input:
- Unique NFT key (_nftKey)
- Borrower address (_borrowerAddress)
- Number of days for which NFT rented (_numberOfDays)
- Start time of NFT renting (_rentalStartTime)
- Amount of ether paid (msg.value)

Requirements:
- msg.sender == _borrowerAddress
- User exists (_borrowerAddress)
- _nftKey should be available for rent
- _rentalStartTime < current time
- _rentalStartTime + _numberOfDays < _dueDate   /*Getter 3 for _dueDate*/
- Sufficient funds paid i.e, msg.value >= paymentRequired
- lenderAddress != _borrowerAddress   /*Getter 3 for lenderAddress*/

Procedure:
- lenderAddress = getLentNftDetails (_nftKey)   /*Getter 3*/
- paymentRequired = _dailyRent * _numberOfDays + _collateral
- if NFT already rented then
      rent FAILS
      return
  end
- lentNft = getUserLentNftDetails (lenderAddress, _nftKey)   /*Getter 5*/
- lentNft.borrowerAddress = _borrowerAddress   /*Update borrower address*/
- _borrowerAddress.rentList.add (_nftKey)
- availableToRentNftList.delete (_nftKey)
- if msg.value > paymentRequired then   /*Return extra amount paid*/
      _borrowerAddress.transfer (msg.value - paymentRequired)
  end
- ERC721 nftCollection = ERC721 (_nftAddress)
- nftCollection.safeTransferFrom (contract address, _borrower Address, _nftId)
- rentalPayment = _dailyRent * _numberOfDays
- lenderAddress.transfer (rentalPayment)

16.5.2.4 Stop lending

This functionality is used when a lender wants to stop lending an NFT already put on lend. This checks that the NFT is not rented by someone else; otherwise, stop lending fails.

Here, the lender takes their NFT back from the contract to stop lending.

Algorithm 4: Stop lending

Input: Unique NFT key (_nftKey)

Requirements:
- User exists (msg.sender)
- nftList.contains (_nftKey)
- availableToRentNftList.contains (_nftKey)
- lentNft.lenderAddress == msg.sender   /*Getter 3 gives lentNft*/
- nftOwner == contract address   /*Getter 1 gives nftOwner*/
- user.lentList.contains (_nftKey)   /*user is msg.sender*/

Procedure:
- nftList.delete (_nftKey)
- availableToRentNftList.delete (_nftKey)
- user.lentList.delete (_nftKey)   /*user is msg.sender*/
- ERC721 nftCollection = ERC721 (_nftAddress)
- nftCollection.safeTransferFrom (contract address, msg.sender, _nftId)

16.5.2.5 Claim collateral

This function is used when the lender wants to claim collateral on the NFT. Also, the lender cannot claim collateral before the rental end time, i.e., before the number of days for which the borrower rented the NFT. Here, as the lender has not received the NFT yet, they claim the respective collateral paid by the borrower in exchange for taking their NFT as rent.

Algorithm 5: Claim collateral

Input: Unique NFT key (_nftKey)
Requirements:
- User exists (msg.sender)
- nftList.contains (_nftKey)
- not (availableToRentNftList.contains (_nftKey))
- rentNft.borrowerAddress != address (0)   /*Getter 4 gives rentNft*/
- User exists (rentNft.borrowerAddress)
- user.lentList.contains (_nftKey)   /*user is msg.sender*/
- _rentalStartTime + _numberOfDays < current time

Procedure:
- lentNft = getUserLentNftDetails (msg.sender, _nftKey)
                                              /*Getter 5*/
- collateral = lentNft.collateral
- nftList.delete (_nftKey)
- rentedNftList.delete (_nftKey)
- borrowerAddress.rentList.delete (_nftKey)
- user.lentList.delete (_nftKey)   /*user is msg.sender*/
- lenderAddress.transfer (collateral)

16.5.2.6 Return NFT

This functionality is used when the borrower wants to return the NFT rented and reclaim their collateral. Here, the NFT is again transferred from the borrower address to either the contract address or the lender address, depending on the time of the due date of the lent NFT. If the NFT due date has passed, the NFT is directly transferred to the lender address; otherwise, it is again transferred to the contract address for others to rent.

Algorithm 6: Return NFT

Input: Unique NFT key (_nftKey)
Requirements:
- User exists (msg.sender)
- nftList.contains (_nftKey)
- not (availableToRentNftList.contains (_nftKey))
- rentNft.borrowerAddress != address (0)   /*Getter 4 gives rentNft*/
- User exists (rentNft.lenderAddress)
- user.rentList.contains (_nftKey)   /*user is msg.sender*/
- Contract address must be approved for Nft transfer

Procedure:
- nft = getNftDetails (_nftKey)   /*Getter 1 */
- lentNft = getUserLentNftDetails (msg.sender, _nftKey) /*Getter 5 */
- collateral = lentNft.collateral
- rentedNftList.delete (_nftKey)
- user.rentList.delete (_nftKey)   /*user is msg.sender*/
- if current time < lentNft.dueDate **then**
  lentNft.borrowerAddress = address (0)
  availableToRentNftList.add (_nftKey)
  ERC721 nftCollection = ERC721 (nft.nftAddress)
  nftCollection.safeTransferFrom   (msg.sender, contract address, nft.nftId)
 else then
  lenderAddress = lentNft.lenderAddress
  lenderAddress.lentList.delete (_nftKey)
  nftList.delete (_nftKey)
  ERC721 nftCollection = ERC721 (nft.nftAddress)
  nftCollection.safeTransferFrom (msg.sender, lenderAddress, nft.nftId)
 end
- user.transfer (collateral)   /*user is msg.sender*/

16.6 GAS GRAPHS

These are the gas graphs for all the functions present in the dApp. Each of the following charts shows units of gas consumed where we consider 1,000,000 units of gas as the limit. To determine the amount of gas to be paid for each of these transactions, it is defined as (gas units) * (gas price per unit), which gives gas fees in Gwei. The current gas price per unit can be fetched from ycharts.com.

Suppose the current gas cost per unit is 60 Gwei.

To calculate the gas price in ether, we can simply use the conversion between Gwei and eth which is defined as

$$1\,\text{eth} = 10^9\,\text{Gwei}$$

We can calculate the gas cost for each of the following functions (Figure 16.7).

Function name	Units of gas	% gas used (limit: 1000000)	Gas cost in ether (units of gas * 60)
addUser()	82899	8.3%	822899*60 = 4973340 Gwei = 0.0049 eth
addNftToLend()	768374	76.8%	768374*60 = 46102440 Gwei = 0.0461 eth
rentNft()	369787	37%	369787*60 = 22187220 Gwei = 0.0221 eth

Function name	Units of gas	% gas used (limit: 1000000)	Gas cost in ether (units of gas * 60)
stopLend()	163592	16.4%	163592*60 = 9815520 Gwei = 0.0098 eth
returnNft()	245478	24.5%	245478*60 = 14728680 Gwei = 0.0147 eth
claimCollateral()	185988	18.6%	185988*60 = 10979280 Gwei = 0.0109 eth

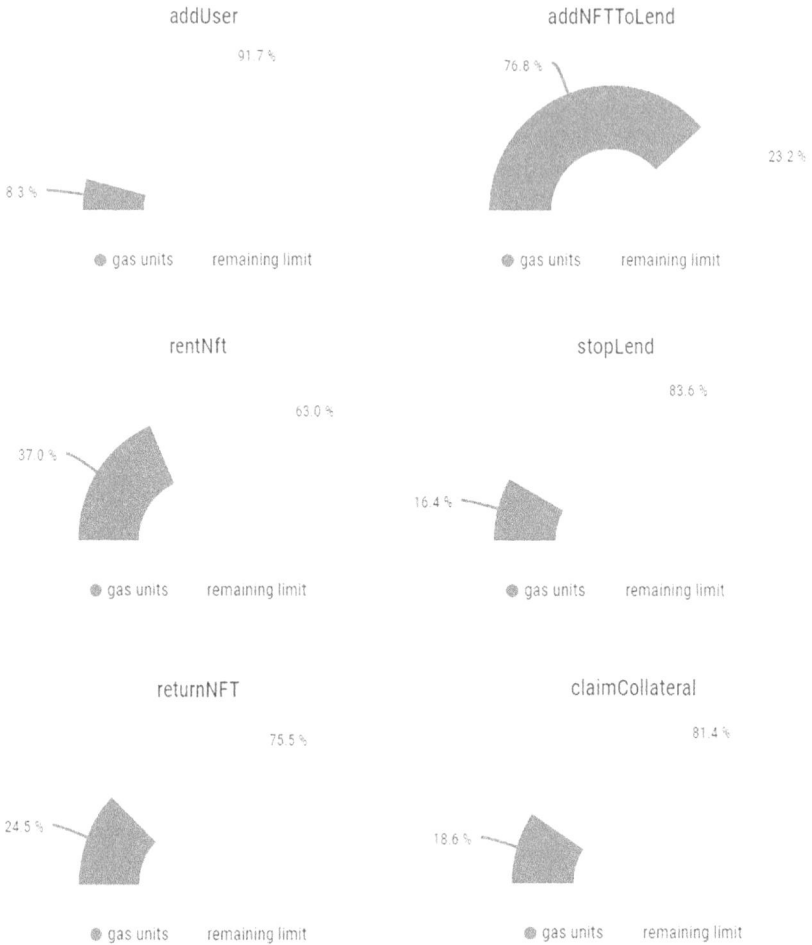

Figure 16.7 Gas graphs.

16.7 REVENUE MODEL

The dApp contract may hold some value as the revenue generated from users in return for its service. The revenue collection may follow various models and forms, a proposed few of which are listed below.

16.7.1 Lending charge

A small amount for lending your NFT to the platform may be charged in order to generate revenue for the dApp. Each time a user adds their NFT to be available to rent, the lending charge may be levied along with the transaction gas fees.

16.7.2 Late fee

In the scenario where the borrower of the NFT defaults on returning the NFT they rented, but the lender has not claimed the collateral yet, we may impose a late fee on the renter while they return the NFT. This charge may also be shared with the lender in appropriate proportion.

16.8 FUTURE SCOPE

Future implementations can add some extended features to make our project more efficient and effective.

16.8.1 Credit score

The credit score defines the reliability of the customer. If the user breaks any predefined criteria, then it will affect their credit score, which will be seen when they rent another NFT.

   On the other hand, if the user's credit score is above a certain level, we can give the user some buffer time if they forget to return the NFT at the rental end time and the lender has not claimed the collateral yet. Our platform thus provides some flexibility for renters of NFTs.

16.8.2 Capped collateral

We can set the maximum collateral up to a specified factor, for example, 1.5x the price of the NFT. A transaction with vague and meaninglessly large collateral will spoil the platform's integrity.

16.8.3 Buy recommendation

This will recommend an NFT that is in demand, and tells how many people bought it earlier.

16.8.4 Rent bidding

People can also bid on popular NFTs and decide who will rent them if they are wanted by multiple users at the same time. The lender will finally accept a bid, and thus the NFT will be rented.

16.9 CONCLUSION

We can, finally, conclude that this project will add another dimension to the use of NFTs. Now people can use other NFTs without purchasing them.

We have developed a prototype that shows how this can be done. We wrote and tested smart contract code to show how NFTs can be rented. We then checked the analytics of how much gas we consumed during the whole process function by function, which is a significant part of the smart contract, to reduce the fees.

REFERENCES

1. Satoshi Nakamoto. *Bitcoin: A Peer-to-Peer Electronic Cash System* [bitcoin.pdf]
2. Qin Wang, Rujia Li, Qi Wang, Shiping Chen. *Non-Fungible Token (NFT): Overview, Evaluation, Opportunities and Challenges* [2105.07447.pdf(arxiv.org)]
3. Foteini Valeonti, Antonis Bikakis, Melissa Terras, Chris Speed, Andrew Hudson-Smith, Konstantinos Chalkias. *Crypto Collectibles, Museum Funding and OpenGLAM: Challenges, Opportunities and the Potential of Non-Fungible Tokens (NFTs)* [mdpi.com]
4. Kamya Pandey. *NFT Renting explained* [jumpstartmag.com]
5. Dragos I. Musan. *NFT.Finance: Leveraging Non-Fungible Tokens* [blockchain test.com]
6. Julia Kagan. *Collateral in Banking* [investopedia.com]
7. Vitalik Buterin. Ethereum Whitepaper [ethereum.org]
8. Why use Moralis? [docs.moralis.io]
9. Openzeppelin ERC721 contracts [openzeppelin.com/contracts]
10. OpenSea Developer Platform [docs.opensea.io]
11. Welcome to Figment Docs [docs.figment.io]
12. Hardhat for beginners: Setup and Testing [hashnode.dev]

Index

Pages in *italics* refer to figures and pages in **bold** refer to tables.

For Product Safety Concerns and Information please contact our EU
representative GPSR@taylorandfrancis.com
Taylor & Francis Verlag GmbH, Kaufingerstraße 24, 80331 München, Germany